# THIN ICE: INUIT TRADITIONS WITHIN A CHANGING ENVIRONMENT

# THIN ICE:
# INUIT TRADITIONS
# WITHIN A CHANGING
# ENVIRONMENT

NICOLE STUCKENBERGER

*with contributions by*

WILLIAM FITZHUGH

AQQALUK LYNGE

KESLER H. WOODWARD

HOOD MUSEUM OF ART, DARTMOUTH COLLEGE

HANOVER, NEW HAMPSHIRE

2007

DISTRIBUTED BY UNIVERSITY PRESS OF NEW ENGLAND

HANOVER AND LONDON

Hood Museum of Art, Dartmouth College
Hanover, New Hampshire 03755
www.hoodmuseum.dartmouth.edu

Distributed by University Press of New England
One Court Street
Lebanon, New Hampshire 03766
www.upne.com

*Thin Ice: Inuit Traditions within a Changing Environment*
Hood Museum of Art, Dartmouth College
Hanover, New Hampshire
January 27–May 13, 2007

This exhibition was organized by the Hood Museum of Art,
Dartmouth College, and generously funded by the John Sloan
Dickey Center for International Understanding, the Evelyn
Stefansson Nef Foundation, the Kane Lodge Foundation, and the
Ray Winfield Smith 1918 Fund and Leon C. 1927, Charles L. 1955,
and Andrew J. 1984 Greenebaum Fund. It was curated by A. Nicole
Stuckenberger, Stefansson Postdoctoral Fellow at the Institute of
Arctic Studies, Dickey Center for International Understanding,
Dartmouth College, as part of International Polar Year.

Edited by Nils Nadeau
Designed by Joanna Bodenweber
Printed by Capital Offset Co., Inc.

Object photography by Jeffrey Nintzel, unless otherwise noted.

ISBN 0-944722-33-4

Cover: Cat. 20. Canada, fish hook with carved seal, late 19th–early
20th centuries, ivory, brass or copper, sinew, fishing line; 29.58.7934.

Title page: Cat. 32. Nunivak Island, walrus mask, collected late
1930s, wood, gold, red, and black paint, sinew, black feathers.
Gift of the Estate of Corey Ford; 169.75.24910.

Song: Cat. 1. West Greenland, Disko Bay, bird spear with throwing
board made by Karl Mathiasen in Sarkak, mid-20th century, wood,
metal, ivory, thread. Gift of Per Jacobi; 160.12.14520.

Table of contents: Cat. 42. Point Hope, Alaska, pair of waterproof
boots, mid-20th century, seal and probably bearded seal leather,
sinew, cord, thread. Gift of Alan Cook, Class of 1960; 163.47.15161.

---

Library of Congress Cataloging-in-Publication Data
Stuckenberger, Anja Nicole.
  Thin ice: Inuit traditions within a changing environment / Nicole
Stuckenberger; with contributions by William Fitzhugh, Aqqaluk
Lynge, Kesler H. Woodward.
     p. cm.
  Exhibition catalog.
  Includes bibliographical references.
  ISBN 0-944722-33-4 (pbk.)
  1. Inuit—Social life and customs—Exhibitions. 2. Inuit—Material
culture—Exhibitions. 3. Inuit art—Exhibitions. 4. Indigenous
peoples—Ecology—Arctic regions—Exhibitions. 5. Arctic regions—
Social life and customs—Exhibitions. 6. Arctic regions—
Environmental conditions—Exhibitions. 7. Hood Museum of Art.
I. Hood Museum of Art. II. Title.
  E99.E7.S83743 2007
  305.897'12—dc22
  2006036773

## Song to the Air Spirit

*Here I stand,*
*Humble, with outstretched arms,*
*For the spirit of the air*
*Lets glorious food sink down to me.*

*Here I stand*
*Surrounded with great joy.*
*For a caribou bull with high antlers*
*Recklessly exposed his flanks to me.*
*—Oh, how I had to crouch*
*In my hide.*

*But, scarcely had I*
*Hastily glimpsed his flanks*
*When my arrow pierced them*
*From shoulder to shoulder.*
*And then, when you, lovely caribou*
*Let the water go*
*Out over the ground*
*As you tumbled down,*
*Well, then I felt surrounded with great joy.*

*Here I stand,*
*Humble, with outstretched arms.*
*For the spirit of the air*
*Lets glorious food sink down to me.*

*Here I stand*
*Surrounded with great joy.*
*And this time it was an old dog seal*
*Starting to blow through his breathing hole.*
*I, little man,*
*Stood upright above it,*
*And with excitement became*
*Quite long of body,*
*Until I drove my harpoon in the beast*
*And tethered it to*
*My harpoon line!*

Igpakuhak, Copper Inuit, Victoria Land,
Central Canadian Arctic (Rasmussen 1932: 138–39)

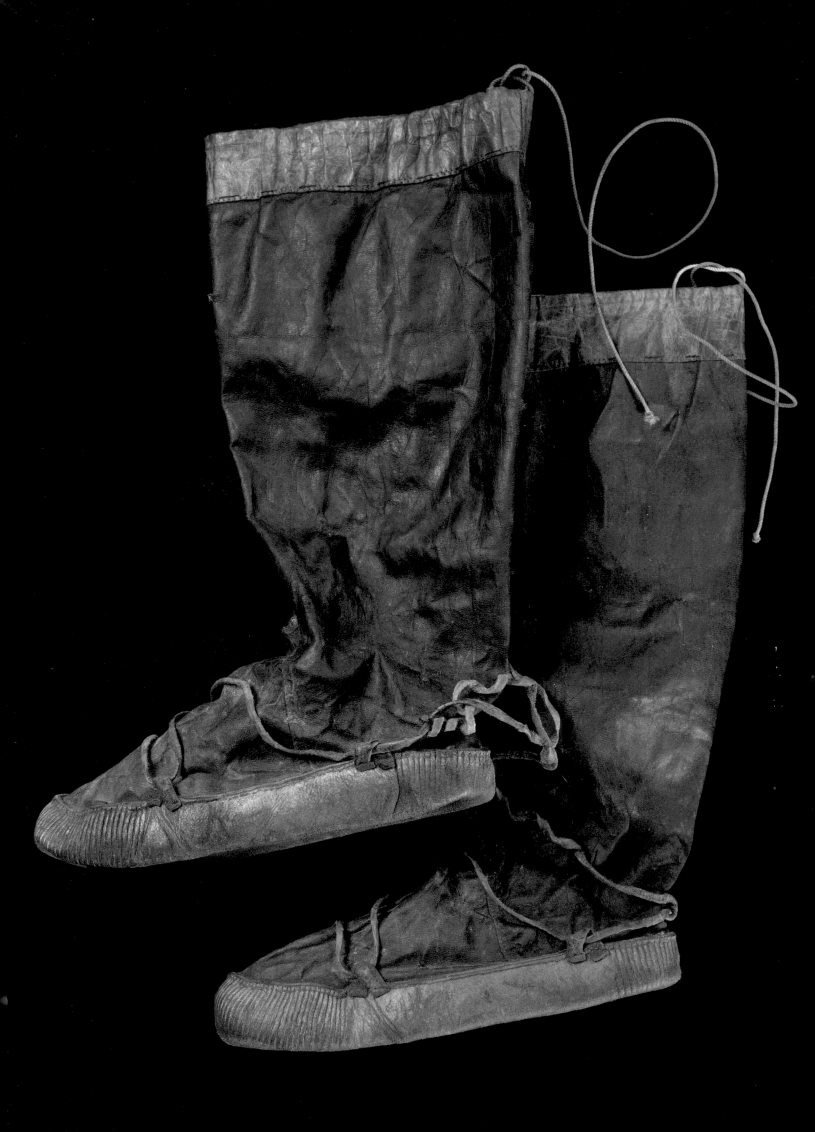

# CONTENTS

8      **Preface and Acknowledgments**
       Brian Kennedy

10     **Foreword: Whose Climate Is Changing?**
       Aqqaluk Lynge

12     **Introduction: Perceptions of Arctic Climate Change**
       Ross A. Virginia, Kenneth S. Yalowitz, and Igor Krupnik

17     **Dartmouth College: A Northern Tradition**
       William Fitzhugh

21     **Arctic, Northwest Coast, and Polar Exploration Collections of Dartmouth College**
       Kesler H. Woodward

29     **Thin Ice: Inuit Life and Climate Change**
       Nicole Stuckenberger

45     **Plates**

57     **Annotated Checklist**
       Nicole Stuckenberger and Erik Lambert

78     **References Cited**

80     **Contributors**

# PREFACE AND ACKNOWLEDGMENTS

The tale of a long-standing collection is often fascinating. Some objects have traveled through many hands before they reach the shelves of a museum storeroom. Others have come via a more direct route. Regardless, all are embedded with the story of their making and with the conditions of their travel from their origin to the present. Most of the objects in the exhibition *Thin Ice: Inuit Traditions within a Changing Environment* have come to rest in an academic and teaching museum at a liberal arts college in northern New England. The Arctic collections were originally housed in the history and anthropology museum at Dartmouth, separate from the fine art collections, but they were all joined together in 1985, when the Hood Museum of Art was built. In an account written by art historian and artist Kesler Woodward in the late 1980s and revised and republished in part here, the absorbing history of these collections is laid out, as well as their later importance in the context of Arctic studies at the College.

The Hood Museum of Art has partnered with the John Sloan Dickey Center for International Understanding and its Institute of Arctic Studies in the development of this first comprehensive exhibition of Dartmouth's Arctic collections, and its accompanying catalogue. It is with great appreciation that we wish to acknowledge the critical roles of Ken Yalowitz, director of the Dickey Center, and Ross Virginia, head of the Institute and Professor of Environmental Studies, for their oversight of this project. Ross Virginia has spearheaded the programs and conferences at the College in relation to the International Polar Year 2007–8 and has also overseen Dartmouth's role in hosting the international gathering for Arctic Science Summit Week in March 2007. We are grateful for his deep involvement and investment in this project's success. We would also like to thank Lenore Grenoble, former Associate Dean of the Faculty for the Humanities as well as Professor of Linguistics and member of the Institute, who worked closely with Professor Virginia to realize this project and bring Nicole Stuckenberger here as the Stefansson Postdoctoral Fellow to work with these collections.

Other members of the Dartmouth community who have assisted with the project include Elmer Harp Jr., Professor of Anthropology Emeritus, an eminent archaeologist who trained Dartmouth students during many field expeditions to the eastern Canadian Arctic and contributed more than three decades of service to the College as Curator of Anthropology in the museum and then as Professor; Jay Satterfield, Director of Rauner Library; Phil Cronenwett, former Director of Rauner Library; Darren Ranco, Assistant Professor of Native American Studies and Environmental Studies; Rob Welsch, Visiting Professor in Anthropology; Sergei Kan, Professor of Anthropology and Native Studies; Hoyt Alverson, Professor of Anthropology; and Heather Carlos, Remote Sensing liaison for the Earth Sciences Department. A number of Dartmouth students contributed to the project, in particular William Blomsted '07, who monitored new publications and projects on Arctic climate change and Inuit society and participated in the development of the exhibition outline; Erik Lambert, MALS '06, who wrote the annotated checklist with Dr. Stuckenberger; and Tiffany Chang '09, who researched material at Rauner Library on scientific exploration of the Arctic. Weyman Lundquist '52, former interim head of the Institute of Arctic Studies on campus, and Kay Taylor also advised Dr. Stuckenberger on the exhibition and her essay for this catalogue. Anne Udry, department administrator for Arctic Studies, provided crucial support for Professor Virginia and Dr. Stuckenberger.

At every stage of the exhibition project, the Hood Museum of Art, the Institute of Arctic Studies, and Dr. Stuckenberger received support from William Fitzhugh '64, Director of the Arctic Studies Center, Smithsonian National Museum of Natural History; Igor Krupnik, Curator of Circumpolar Ethnology at the Arctic Studies Center; and Barbara Stauffer, Program Manager and Exhibit Developer at the Arctic Studies Center. Drs. Fitzhugh and Krupnik have visited Dartmouth and taught with the Arctic collections here; they also contributed to this catalogue. We would like to thank Deborah Nichols, Chair of the Anthropology Department at Dartmouth, for her role in bringing them to campus on more than one occasion. Drs. Krupnik, Fitzhugh, and Ms. Stauffer also organized an inspirational exhibition titled *Arctic: A Friend Acting Strangely* at the Smithsonian Natural History Museum in 2006.

In addition, Nicole Stuckenberger would like to thank the following scholars and curators for their advice and feedback: Jean-Loup Rousselot, Staatliches Museum fur Völkerkunde, Munich;

Cunera Buijs, Rijksmuseum voor Volkenkunde, Leiden (IPY partner); Geert Mommersteeg, Brenda Oude-Breuil, Jacobijn Olthoff, and Yvon van der Pijl, Utrecht University; Rieke Lenders, University of Amsterdam/Russian Study Center; Christien Klaufus, Delft University/OTB; Sonja Leferink, independent scholar and anthropologist; Jarich Oosten and Barbara Miller from Leiden University; Cor Remie, Nijmegen University (Netherlands); Wim Rasing, Research Group Circumpolar Cultures (Netherlands); Marianne Stenbaek and Ludger Mueller-Wille, McGill University, Montreal; Frederic Laugrand, Bernard Saladin d'Anglure, and Louis-Jacques Dorais, Laval University, Quebec; Norman Vorano, Canadian Museum of Civilization; Cynthia Jones, retired Director/Curator of the Sheldon Museum, Haines, Alaska; and Claudio Aporta, Carleton University, Ontario.

We are grateful to Victor Rabinovitch of the Canadian Museum of Civilization and Debra Moore of the Hudson Bay Archives for important loans to the exhibition, and to Cherry Alexander of Arctic Photo for images for the exhibition.

We further thank Mr. Aqqaluk Lynge for his insightful statement to open this catalogue. Mr. Lynge, an Arctic poet and statesman, has served tirelessly to advance the rights of Indigenous peoples both in Greenland and globally as a member of the United Nations Permanent Forum on Indigenous Issues and as President of the Inuit Circumpolar Council, Greenland.

At the Hood Museum of Art, numerous people contributed to the realization of this exhibition. Katherine Hart, Associate Director and Barbara C. and Harvey P. Hood 1918 Curator of Academic Programming, was the project's chief coordinator, giving it shape and style with her customary dedication; Barbara Thompson, Curator of African, Oceanic, and Native American Collections, made important contributions to the exhibition catalogue manuscript; Nils Nadeau, Publications and Web Manager, edited and oversaw the exhibition catalogue; Kris Bergquist, School and Family Programs Coordinator, was the Education Department liaison for the exhibition; Kellen Haak, Collections Manager and Registrar, and Deborah Haynes, Database Manager, provided advice and expertise on the museum's collections; Patrick Dunfey, Exhibitions Designer/Preparations Supervisor, designed a complex and handsome exhibition installation; John Reynolds and Matt Zayatz, Lead Preparator and Preparator, worked on all aspects of planning and implementing the installation; Juliette Bianco, Assistant Director, and Nancy McLain, Business Manager, oversaw its budgets; Kathleen O'Malley, Associate Registrar, arranged for photography of a collection that had previously been underdocumented; Jeffrey Nintzel photographed both Hood and Rauner Library objects; and Sharon Reed, Public Relations Coordinator, publicized the exhibition to the regional and national press. Sally Eshleman and Carole Cutler from the College's Foundation Relations Office gave invaluable help with development issues. Joanna Bodenweber designed the handsome catalogue, which is distributed by University Press of New England.

Nicole Stuckenberger arrived at Dartmouth two years ago to undertake this ambitious project and has been a dedicated colleague and scholar throughout her work on Thin Ice: Inuit Traditions within a Changing Environment. We have benefited immensely from her expertise on the subject of Inuit society and culture, and her efforts have produced a marvelous exhibition and catalogue on an important subject. From the first she felt it was imperative to see the Arctic collections as part of a larger picture involving both Inuit tradition and cosmology, which framed these objects within the pressing issue of climate change in that region and its effects on Inuit existence today.

As with any project of this nature, it takes true believers and enthusiastic supporters to get it off the ground. Evelyn Stefansson Nef, along with her late husband, Vilhjalmur Stefansson, has been one of the movers and shakers behind Arctic studies at the College, and through the Evelyn Stefansson Nef Foundation she provided important funding for both this exhibition and for the International Polar Year activities as a whole. Major funding for Thin Ice also came through the John Sloan Dickey Endowment for International Understanding and the Kane Lodge Foundation, Inc., named for Elisha Kent Kane, one of the first American Arctic explorers. The Hood Museum of Art's endowment funding from the Ray Winfield Smith 1918 Fund and the Leon C. 1927, Charles L. 1955, and Andrew J. 1984 Greenebaum Fund makes such quality exhibitions possible.

We are gratified to have participated in this project and hope that it will add considerably to knowledge of and interest in Arctic Studies within our changing world environment today.

Brian Kennedy
Director, Hood Museum of Art

# FOREWORD: WHOSE CLIMATE IS CHANGING?

We are all in this together. Yet our perceptions are different. This is the challenge.

Climate change is now part of the vocabulary of almost everyone: industry, Western science, governments, Inuit, and others. But our respective relationships to the land, sea, animals, and sea mammals differ. And what each of us sees in the now changing *sila* and what it means for each of us differs.

How can we use our different perceptions of change in *sila* and, perhaps more importantly, how can we reconcile our different and sometimes competing interests when it comes to effectively dealing with climate change?

Without a doubt, the Arctic will be—and already is—one of the hardest hit places in the world. And those that will be most affected are my people, the Inuit.

With industry leaders salivating at the thought of an Arctic sea route opening up to them, Inuit are concerned about what this means for their lands and seas. With governments changing their perceptions on what needs to be done—one year a government signs on to Kyoto, the next year another government backs away from its commitment—Inuit fear time is slipping by as our ice keeps on thinning and melting.

The English looked for the Northwest Passage, and the Spanish, the Strait of Anián. Since the early sixteenth century, others have been looking for a commercial route through our backyard, and it seems that through climate change they have found one. The cultural, environmental, economic, and spiritual impact of this discovery on those living in the Arctic will be monumental. Inuit are inviting those with different perceptions of *sila* into a dialogue so that they can avoid, mitigate, or deal head on with what is now happening, and what is about to come.

Our hunters, with Inuit science guiding them, were the first to take note that something was radically different. They came to us with reports of thinning ice, disappearing ice floes, changing animal migration patterns, and eroding shores. Their traditional knowledge, which they received from their grandparents, who in turn received it from their grandparents before them, had given them an understanding that animal migration patterns change, as does the climate. But something

was different, they told us. They could no longer rely on their hunter knowledge in the same way.

The Inuit Circumpolar Council and other Inuit organizations brought these reports to the attention of scientists, the media, industry, and governments. Each responded in various ways and in varying degrees. Rather than turning off the greenhouse taps, industry saw a commercial opportunity. The media saw a great story. Governments scrambled, looking for solutions to the problem, or projects for the potential opportunities created. Western scientists saw interesting research.

As Inuit face this uncertain future, they are asking for dialogue and partnerships with all those that view the changing climate from their own vantage point. We believe, in fact, that there are overlaps. We have different perceptions, but when we look closer, we see intersecting interests. It is on these points of intersection that we must focus in order to avoid the potential cultural and spiritual calamity. The partnerships created among Inuit, scientists, and governments at the Arctic Council are a good start. These partnerships must be strengthened. Inuit are so often left out of the research process, and this must never happen with respect to our changing climate. We must reach out to industry, and industry to us. They must be told that Inuit will not tolerate ships in their seas if they end up destroying us as a people. But they should also be aware that we want dialogue and partnerships, and that we believe solutions may somehow be found.

Our *sila* sustains us and Inuit are resolved to keep it that way for millennia to come. As our sea ice thins, and as our ice caps melt, Inuit will pursue solutions to these challenges through partnerships with those who perceive *sila* differently from us. We are in this together, and it is that knowledge that will guide us.

Aqqaluk Lynge
*President, Inuit Circumpolar Council, Greenland*

Cat. 25. Greenland, miniature one-man kayak, fully
equipped, collected 1950, sealskin, gut, ivory, wood, steel.
Gift of Peter S. Dow; 50.14.12412.

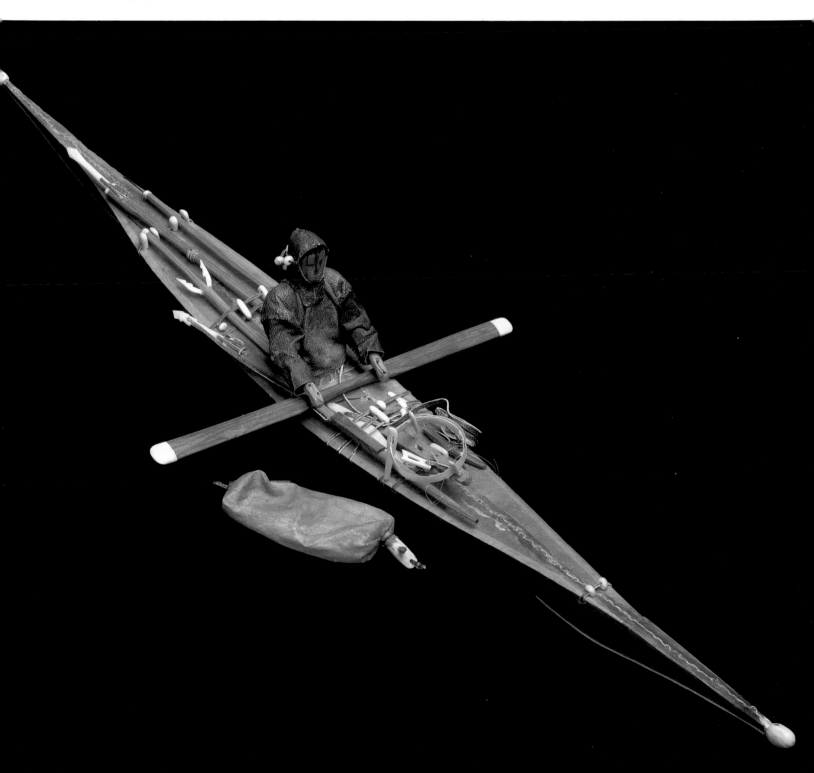

# INTRODUCTION: PERCEPTIONS OF ARCTIC CLIMATE CHANGE

*Climate is what we expect, weather is what we get.*
Attributed to Mark Twain

*In Alaska, the beaches are slumping so much, people are having to move houses. In Tuktoyaktuk, the land is starting to go under water. The glaciers are melting and the permafrost is melting. There are new species of birds and fish and insects showing up. The Arctic is a barometer for the health of the world. If you want to know how healthy the world is, come to the Arctic and feel its pulse.*

Sheila Watt-Cloutier, former chair of the Inuit Circumpolar Conference (in DeNeen L. Brown, "Poisons from Afar Threaten Arctic Mothers, Traditions," *Washington Post* Foreign Service, April 11, 2004)

The earth's polar regions have been the subject of three major research initiatives called "international polar years" (IPY). Beginning with the first IPY in 1882–83, these events have shared the goal of advancing basic scientific knowledge of the geography and geophysical processes of these remote lands and oceans via global conferences and research initiatives at thirty- to fifty-year intervals. International polar year events have always captured the imagination of the public, yet the polar regions remain a distant, and disconnected, realm for most people. The global science community is set to begin another IPY in 2007–8 with a special sense of urgency: simply put, the polar regions are a critical part of the earth's climate system, which is now undergoing rapid change in response to human activities.

The 2007–8 events extend beyond basic studies in the geophysical and biological sciences to focus on global climate change and research to advance our understanding of the human dimensions of a shifting Arctic environment. Both current climate science and the observations of polar residents warn us that the Arctic region (variously defined as the northern polar region, the lands and ocean north of latitude 66.5 degrees, and the circumpolar region extending from the northern limit of tree growth to the North Pole) is experiencing unprecedented warming and is acting as a "canary in the coal mine" to warn of serious environmental problems extending throughout the global ecosystem. As Sheila Watt-Cloutier admonishes us, the state of the Arctic ultimately reflects the health of the Earth.

The exhibition *Thin Ice: Inuit Traditions within a Changing Environment* explores various perceptions of change in polar weather and climate and how they are shaped by the culture behind them. In the Western scientific tradition, we see "weather" as the changing atmospheric conditions of temperature, precipitation, and wind over short periods (days and weeks). If we "average" these conditions over longer periods (years and decades or more), then we speak of "climate." Distinctions between weather and climate arise in the West when modern instruments are used to precisely quantify change over time. But how are these distinctions—and the lessons to be learned from them—perceived and recorded by people who have lived in the Arctic for generations without the use of satellites, thermometers, or computer models? *Thin Ice* explores the lives of the Inuit people of the Arctic and their intimate relation to ice, weather, climate, and nature, the many manifestations of the Inuit concept of *sila* (universe, weather).

*Thin Ice* is a contribution by Dartmouth College and its colleagues to the international polar year as part of *Project 160: Arctic Change: An Interdisciplinary Dialog between the Academy, Northern Peoples, and Policy Makers*. The exhibition is a collaboration between Dartmouth's Hood Museum of Art, the Institute of Arctic Studies within the Dickey Center for International Understanding, the Rauner Special Collections Library, and the Arctic Studies Center of the Smithsonian National Museum of Natural History. Located in Hanover, New Hampshire (roughly halfway between the equator and the North Pole), Dartmouth has a distinguished tradition in Arctic research reaching back to Vilhjalmur Stefansson (1879–1962), the famous Arctic explorer, scholar, and founder of Dartmouth's Northern Studies program. Stefansson's long career in Arctic studies began in earnest a century ago with his first expedition in 1906–7 to the McKenzie River Delta region of the Canadian north. Stefansson became known for adopting Inuit local ecological knowledge and ways of traveling and living in the north. "Stef" wrote prolifically of the "Friendly Arctic," a region supporting uniquely adapted people and containing abundant natural resources. He foresaw many of the social and political changes that have since swept the Arctic, but he could never have anticipated that the Arctic sea ice he once lived on would be so much thinner, or that the frozen permafrost lands across which he traveled would be melting.

## Perspectives on Change: Western Science, Indigenous Knowledge, and Policy

"Earthrise," the famous satellite image of Earth taken by the Apollo 8 astronauts in 1968 as they came from behind the moon after achieving lunar orbit, transformed scientific thinking about the interconnectedness of land, sea, and atmosphere. A major international effort titled the Arctic Climate Impact Assessment (ACIA 2004) gives the scientific consensus on climate change for this region. It reports that Arctic temperatures are now rising at nearly twice the rate seen for the rest of the world, with resultant reductions in the extent and thickness of sea ice and the melting of frozen soils. Continued warming, caused primarily by greenhouse gas emissions, is projected to contribute an additional eight to fourteen degrees Fahrenheit over the next one hundred years. As the Arctic warms, the extent of the sea ice covering the Arctic Ocean has declined over the past thirty years by an area roughly equal to Texas and Arizona combined, and the ice that does remain is now thinner by six feet in many locations (fig. 1). It is predicted that by the end of the century most of the Arctic Ocean will remain ice-free during the summer months. The melting of previously frozen soil also disrupts transportation and damages buildings. The duration that vehicles can drive over frozen Arctic tundra ground for oil exploration has declined from about 200 days in 1970 to only 125 days in 2000. Global climate change is expected to increase the frequency and intensity of extreme weather events, such as hurricanes and droughts. In the Arctic, reductions in the coverage of near-shore ice have allowed extreme storms and wave action to erode shorelines, forcing many subsistence-based communities to relocate inland or otherwise act to mitigate the emerging threat.

Climate science, anthropology, history, and other fields in the sciences, social sciences, and humanities are joining in new interdisciplinary partnerships to understand the forces behind climate change, and its human dimensions as well. For example, we can gain knowledge by comparing historical accounts, early observational records, and the oral descriptions of Inuit elders. The early written accounts of explorers like Stefansson, ship captains, trappers, and missionaries provide ample information on weather, the extent and duration of Arctic ice cover, the seasonal activities of animals, and vegetation dynamics against which contemporary observations can be compared (fig. 2). There is a wealth of information on local change in ecosystems, ice conditions, and the weather taken from the oral tradition and material culture of polar indigenous people such as the Inuit. Often referred to as Traditional Ecological Knowledge (TEK), this way of knowing provides an understanding of the environment that is much more holistic and closely

Observed sea ice, September 1979

Observed sea ice, September 2003

Fig. 1. Extent of Arctic sea ice in September 1979 and September 2003, constructed from satellite data. September is typically the least icy month, and 1979 was the first year satellite data of this type became available (ACIA 2004, http://amap.no/acia/Files/ObsSeaIceNASA1979_03_150.jpg).

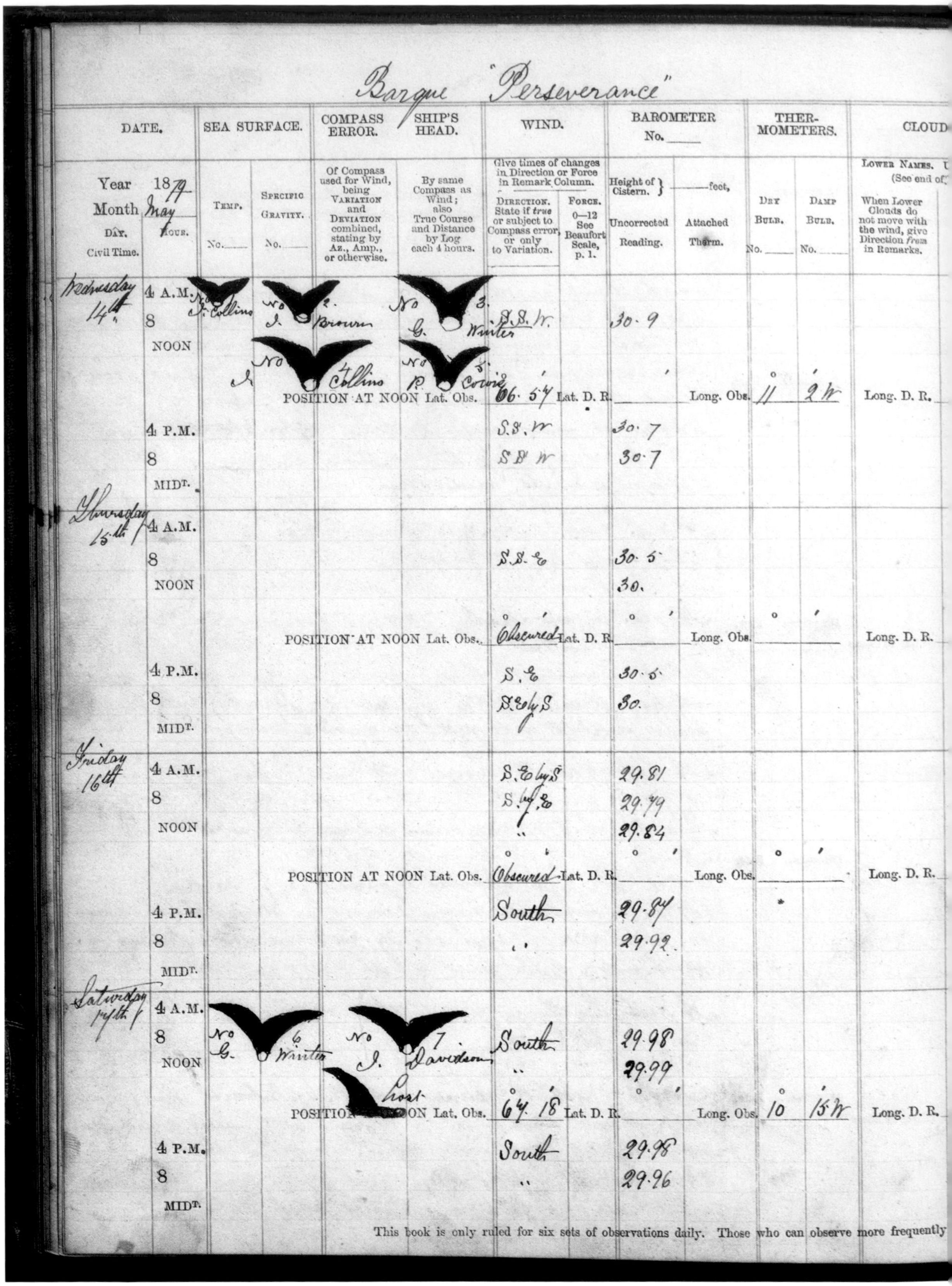

14

Fig. 2. An entry from the logbook (cat. 66) of the *Persever-ence,* a whaling ship on voyages from Peterhead, Scotland, to Baffin Island during 1877–80. A log entry from May 1879 shows the daily weather observations of wind, baro-metric pressure, clouds, sea surface, and so on. The suc-cessful capture of a whale is marked by a sketch showing both dorsal fins; a harpoon strike without a capture is shown by a single dorsal fin (Stefansson Collection on Polar Exploration, Rauner Special Collections Library, Dartmouth College Library; Stef MSS-121).

tied to Inuit life than data derived from the methods and instruments of Western science.

The current rapid Arctic climate changes that we are experiencing will forever alter the balance between ecosystem services provided by the ocean, animal harvest, forestry, agriculture, mineral resource extraction, and other important economic activities such as tourism. The potential for people to be engaged in these activities will likely be shaped by the northward-moving boundary between Arctic forest and treeless tundra, by the increasing abundance of more southerly species, and by the decline of some polar species (such as the polar bear) as sea ice thins or disappears. Subsistence hunting and herding, key elements of Arctic life to this day, may also be disrupted. New and potentially serious political, economic, and strategic issues may result from these Arctic ecosystem changes. Thinning sea ice might mean that the exploitation of oil, gas, and mineral resources will be enhanced; year-round shipping lanes will open; and the continental shelf claims of several nations will be affected. Ideally, this would all lead to more cooperation and multilateralism in dealing with Arctic issues. Unfortunately, the current governance systems across the Arctic are neither regulatory nor vested with the policy-making authority necessary to deal with possible negative environmental and social consequences. A key challenge for the academic community, therefore, is to convince policymakers to focus on these issues multilaterally and cooperatively, and with the full involvement of native Arctic residents, to avoid ad hoc national reactions that could heighten tensions.

Can the consequences of Arctic climate change be anticipated so as to spur the implementation of new policies that manage Arctic resources in a more sustainable way for the benefit of all northern inhabitants? Our collective ability to adapt to Arctic climate change, or to mitigate its effects, will depend upon a productive collaboration between northern peoples, the scientific community, and policymakers from the local to the international levels. Without a dialogue involving indigenous perspectives and timely policy actions, the future of the "Arctic" and perhaps the entire planet may truly be on thin ice.

Ross A. Virginia
*Director, Dickey Center Institute of Arctic Studies, Professor of Environmental Studies, Dartmouth College*

Kenneth S. Yalowitz, Ambassador (Ret.)
*Director, Dickey Center for International Understanding, Adjunct Professor of Government, Dartmouth College*

Igor Krupnik
*Curator, Circumpolar Ethnology, Arctic Studies Center, National Museum of Natural History, Smithsonian Institution*

# Dartmouth College: A Northern Tradition

WILLIAM FITZHUGH

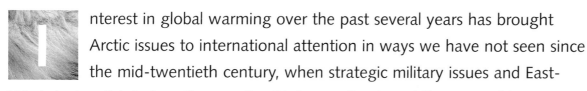 nterest in global warming over the past several years has brought Arctic issues to international attention in ways we have not seen since the mid-twentieth century, when strategic military issues and East-West rivalry dictated northern policy. Today we live in a different world. Former rivals travel in each other's Arctic territories, share joint expeditions, co-publish scientific reports, and have access to each other's public and scientific media. We are now very much aware that we share a single Arctic.

This new climate of cooperation offers an unprecedented and timely opportunity for study and research on climate change and its environmental and human impacts. The most important outcome of this convergence of political will and scientific need has been the collective call to action exhibited by the establishment of the 2007–8 International Polar Year and its various research and public awareness programs, all of which encourage international cooperation, a focus on global Arctic priorities, and, for the first time ever, the active participation of indigenous northern societies.

Cat. 45. Caribou parka with caribou pants, canvas over-pants, and caribou mittens, 1913–18, from Canadian Arctic Expedition: possessions of James Crawford, Stefansson Collection on Polar Exploration, Rauner Special Collections Library, Dartmouth College Library; Stef-Realia 213.

Part of the present international polar year's agenda is geared toward strengthening institutions and educational training capabilities. It is here that Dartmouth College can make an important contribution through its unique position as a primarily undergraduate institution with a strong tradition of northern research and training, excellent libraries and facilities, Arctic archives, and museum collections, and a commitment to attracting students from northern cultures and lands. A unique feature of this tradition has been its broad interdisciplinary approach and its international networks, early on with a Canadian Arctic and Alaskan focus, and in recent years with strong programs in studies of the Soviet and Russian north as well. While remaining a relatively small school, Dartmouth continues to have a large impact, not only through its Institute of Arctic Studies and Environmental Studies and Russian programs but also through its research and teaching in anthropology, earth sciences, geography, government, linguistics, and other disciplines. Every year these programs capture the minds and hearts of a few students who go on to graduate school and ultimately specialize in northern or polar studies. And for the many others who do not, broad-based environmental, humanities, and cultural studies classes attract students who carry their knowledge of the North forward into many other walks of life.

For all of these reasons, *Thin Ice: Inuit Traditions within a Changing Environment* is an important contribution by Dartmouth to IPY 2007–8 and to the College's evolving northern portfolio. Prominently featured in this exhibition, which coincides with the centennial of Stefansson's first expedition to the northern polar regions, are the Arctic collections of the Hood Museum of Art and materials from the explorer's own extensive collection of materials and manuscripts. As an anthropology graduate in 1964, I had occasion to use both collections, and they were instrumental in my decision to specialize in archaeology and cultural studies in the Arctic and Subarctic. The Stefansson Collection came with a bonus, for by studying it at Dartmouth in the early 1960s I had a chance not only to browse among the lives and exploits of Arctic explorers but to meet some of them as visiting scholars and researchers. Where else could an undergraduate hope to meet in person such diverse polar luminaries as a kilted Farley Mowat, breezing in to consult Stef's papers for one of his spellbinding northern yarns, or Alan Cook, the erudite bibliographer of the Canadian Arctic?

I had similar experiences in the old College Museum, where most of my anthropology courses were taught by Bob McKennan, Elmer Harp, and Al Whiting. Although the museum exhibit case I prepared for Professor Whiting's 1963 museum methods class was on Arabian Bedouins (the museum even then contained a wealth of art and cultural materials from around the world, donated by alumni or the result of staff field collecting), it gave me my first training in museum research and exhibit techniques. Ten years later, after becoming a curator at the Smithsonian's Anthropology Department, I returned to Dartmouth and was surprised to see the case still up! I like to think it was the excellent presentation, but as a professional curator, I suspect it may have been the museum version of "a bird in the hand is worth two in the bush." Since 1985, the Hood has done it differently, combining art and anthropology in a more lively fashion. I am glad to see that the College's fine Arctic collections, with especially superb materials representing art and ethnographic materials from Native Alaska, Canadian, and Greenlandic cultures, are being used by anthropology and Arctic studies students. For the past few years I and my colleagues from the Smithsonian's Arctic Studies Center have used these materials and the Stefansson Archives for classes we have taught at the College as well.

This exhibition and its catalogue are truly a historic occasion for Dartmouth's northern program, representing the first time that a substantial number of the Hood's Arctic and Subarctic anthropological objects have been brought out from storage and introduced in an exhibition format, as well as published in a fine, contextualized catalogue that illustrates the relevance of historical collections to modern-day issues. Many of the objects exhibited and illustrated here were collected by Dartmouth students, faculty, and friends as early as the late nineteenth and early twentieth centuries, when the north was a very different place than it is today, as documented by Nicole Stuckenberger's essay. Although underutilized for many years (a not uncommon situation for university museum collections), Dartmouth's Arctic collections are important because of their geographic scope—representing virtually all areas of the circumpolar Arctic—their broad chronological sweep from early exploration to modern times, and their diverse subject matter, touching upon comparative technology, dress, and art, as well as social and religious concepts.

Anthropological collections derive special value from their documentation and collecting contexts, without which they lose much of their research and educational value. Many of the objects presented here are everyday items—skin scrapers for preparing hides, fishhooks and sinkers, boots and mitts—yet through the miracle of conservation and modern museum care, lighting, good photography, and careful research, these emissaries from the past can return to provide insight and inspiration into changing environmental conditions and ways of life. Coming from different places and times, they reveal ways of thinking and behaving under conditions that mostly no longer exist, therefore serving as signposts to the past as well as the future. Today, these materials are avidly sought after by scholars and the native descendants of those who created them for their artistic value, craft, and representation of adaptations to a wide range of environmental conditions and cultures. I am particularly pleased to see Dartmouth's Arctic collections reemerge at such an important time in our evolving understanding of Arctic regions and peoples.

Since 1985, courses, special programs, and exhibitions continue to provide opportunities for Dartmouth faculty, students, and the wider community to pursue research, study, and interests in the North. The Dickey Center for International Understanding, the home base for the Institute of Arctic Studies, sponsors research and events that link Arctic issues and politics with the greater world community. The opening of the Russian Arctic has made a huge area of the Arctic region accessible for the first time in generations, so now we can examine its deep geological, climatic, environmental, and human histories in the context of both circumpolar and global issues. Dartmouth, with its recognized tradition of Arctic scholarship and educational experience, is well-positioned to respond creatively to the challenges ahead, and I hope it will continue to strengthen its interdisciplinary Arctic focus. *Thin Ice* looks back to what we can learn from anthropological research, traditional Inuit knowledge, and the pioneering contributions of scholars and explorers like Vilhjalmur Stefansson. At the same time, it challenges us to consider the increasingly rapid pace of climate change and its influence on the Arctic people, a timely enterprise given this international polar year. By combining the efforts of its physical, social, and cultural faculties in the name of its long-term commitment to native studies, Dartmouth can continue to inquire for, lead, and serve society as we enter the climatically uncharted (and increasingly ice-free) Arctic waters ahead.

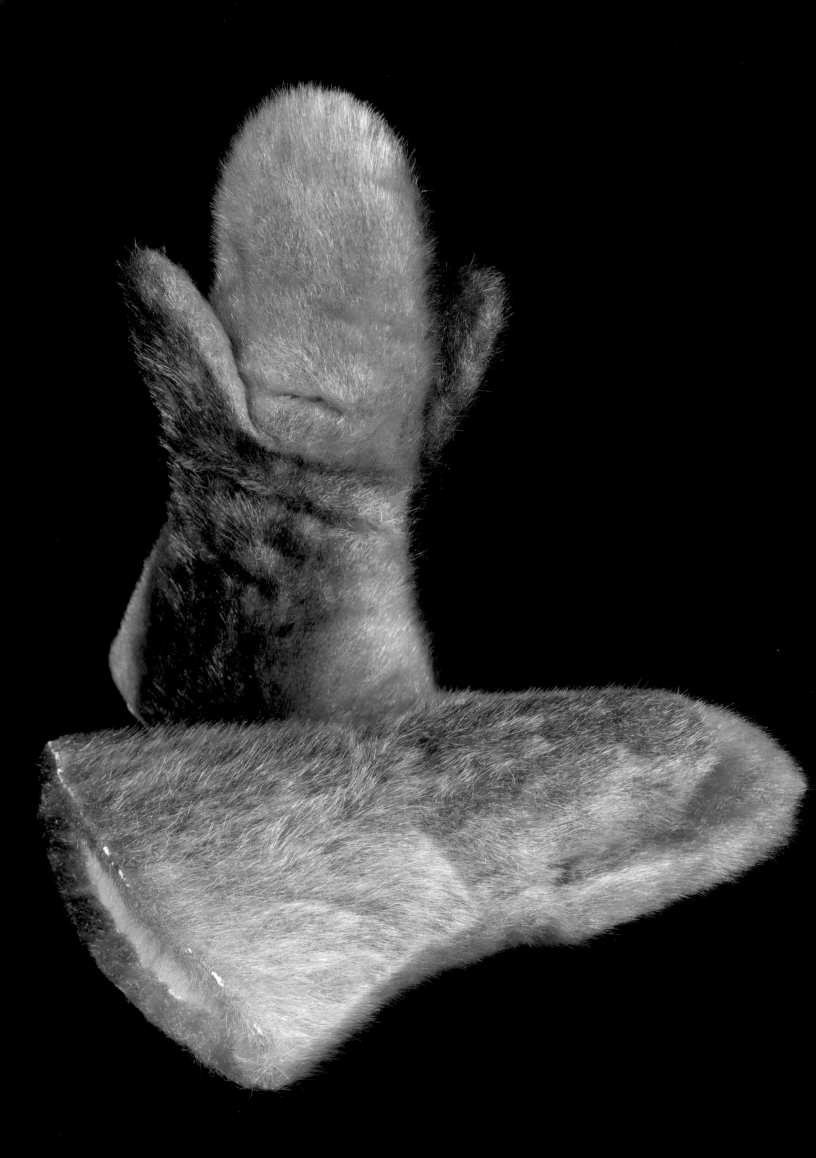

# Arctic, Northwest Coast, and Polar Exploration Collections of Dartmouth College

KESLER H. WOODWARD

artmouth College has in the Hood Museum of Art and Rauner Special Collections Library a substantial number of objects relating to the art and material culture of Canada and the circumpolar North. The Arctic collection at the Hood Museum of Art numbers more than three thousand items. Rauner Special Collections Library holds the Stefansson Collection on Polar Exploration, where a number of interesting paintings, watercolors, drawings, prints, and sketchbooks are kept, together with their accompanying manuscript material. It should be noted that there are also about fifteen thousand photographs in the Stefansson Collection. While these are an important resource, and are in many cases of considerable visual as well as documentary interest, they have not been considered in this review.

Although Dartmouth College's long association with northern and Arctic research is well known, and the impressive Stefansson Collection of books and manuscripts has been widely recognized and used for many years, there has been less awareness of the College's Inuit collections. The establishment of the Institute of Arctic Studies in the late 1980s reinvigorated and nourished interest

A version of this essay was originally published in issue no. 1 of *Northern Notes*, an occasional publication of the John Sloan Dickey Endowment for International Understanding at Dartmouth College, in November 1989.[1]

Cat. 41. West Greenland, pair of sealskin mittens, double-thumbed, 1897, sealskin and fur lining (probably dog fur). Gift of Mrs. William Stickney, wife of William Stickney, Class of 1900; 164.21.15432.

in these collections, and it is as a part of that ongoing effort, and as a service to those who study northern art and ethnographic material, that this review of the College's Inuit art was undertaken at the time of the Institute's founding.

Much of Dartmouth's northern material came as the direct or indirect result of the remarkable period of northern activity at the College in the 1950s and 1960s.[2] The acquisition of the Stefansson Collection in 1952,[3] the presence of Vilhjalmur and Evelyn Stefansson on campus throughout the 1950s,[4] and the staunch support for Canadian and Arctic endeavors at Dartmouth on the part of then-President John Sloan Dickey all combined to make the College a leader in northern activity during this period.[5] Donations, including several sizable groups of objects, account for a major portion of the collection. There have been occasional purchases of single fine objects or groups of works to augment these holdings, but it is a testimony to the strength of the northern impetus at the College, and to the loyalty and affection of the College's students, faculty, and friends, that these collections are so substantial in number and include so many fine objects.

## The Arctic and Northwest Coast Collections of the Hood Museum of Art

Dartmouth's earliest-collected and earliest-accessioned northern ethnographic material is probably a group of Aleut and Koniag objects, several of them from Kodiak Island and vicinity, that came to the College in the late nineteenth century. The group includes gut (seal intestine) bags, a gut parka, a whaling spear and harpoon, a fox trap, and several kayak models. Although the nineteenth-century museum records are unclear, it seems that some of that material may have come to Dartmouth through the Captain and Mrs. Worthen Hall donation, a bequest from their daughter in 1887. Worthen Hall was captain of a whaling vessel from 1837 to 1855, and his daughter willed his collection of "shells, corals, and curiosities"[6] to the old Dartmouth College Museum. Additional material attributed to the Worthen Hall donation includes three whale teeth with scrimshaw, two gut caps, another gut parka, and a pair of boots, all identified simply as "Alaskan" in the Hood Museum of Art records.

Even more interesting among the early accessions is the gift of Mrs. Margaret Kimberly, a sizable donation of work from a number of cultures that includes a substantial amount of fine material from the Northwest Coast area. Mrs. Kimberly's uncle, General John Hewston, had been a friend of a sea captain active in the Pacific between about 1820 and 1860, and the donated material is thought to have been acquired by General Hewston from the captain. Accessioned in 1922, the collection of twenty-five Haida, Kwakiutl, and Tsimshian objects includes rattles, wood and horn spoons, carved wood fish hooks, a spruce root hat, a shaman's charm, and an argillite pipe.

The real foundations of the Hood Museum of Art's northern acquisitions are several large donations and bequests, perhaps most important among them the Frank C. and Clara G. Churchill Collection, received as a bequest upon Mrs. Churchill's death in 1946. Frank C. Churchill was a prominent local (Lebanon, New Hampshire) civic leader and manufacturer who served as Special Inspector in the United States Indian Service from 1899 to 1909. During that period, he and his wife crossed the continent ten times and visited more than one hundred Indian groups, from which Mrs. Churchill collected artifacts.[7] In 1905, Churchill was appointed special emissary to Alaska to investigate Inuit schools and the government-sponsored reindeer domestication program. The couple spent three months traveling more than ten thousand miles along the coasts of Alaska and the islands of the Bering Sea, largely aboard the U.S. revenue cutter *Bear*. Among the approximately 1,400 native North American objects in the Churchill bequest are 187 Alaskan Inuit artifacts and forty-five Northwest Coast objects collected during that trip.

The Churchill Collection includes the travel journals of Clara Churchill, which contain site descriptions, purchase prices, and other information about the objects collected.[8] Ivory carvings, wood and sheep horn spoons, tools, projectiles, models, games, and clothing are all to be found in the collection, but the most numerous artifacts are baskets, including northern items from Point Clarence, Golovnin Bay, Kotzebue, Attu, Atka, Yakutat, Sitka, and British Columbia, and several relatively rare Pacific Inuit spruce root baskets.[9]

Former Hood Museum of Art Curator Tamara Northern commented on the strengths and weaknesses of the Churchill Collection: "The character of the Churchill material resembles that of other smaller ethnographic collections at Dartmouth in that it is strong in the types of objects that were readily accessible to the public. . . . Around the turn of the century, when the Churchills collected, these articles were available for purchase, and many were made for sale and bought by Mrs. Churchill directly from the artisan. Articles of ritual and religious significance are almost totally absent from the Churchill Collection."[10]

Another important collection in the Hood Museum of Art consists of several groups of material collected by Axel Rasmussen while he was a superintendent of schools in southeast Alaska in the late nineteenth century. These came to the College between 1959 and 1966, largely through donations from Doris Meltzer and the Meltzer Gallery in New York. Much of the outstanding material Rasmussen collected was eventually to form the important Rasmussen Collection of Northwest Coast Art at the Portland (Oregon) Art Museum.[11] The pieces at Dartmouth are not the showiest of the objects collected by Rasmussen, but they are well-documented everyday items from a quite early period, including a number of nineteenth-century Northwest Coast baskets, raw materials for baskets and woven items, wooden boxes, knives, and adzes.

During the l960s, Sherman P. and Anne L. Haight of New York City donated to the College another of the larger collections of northern material now at the museum. This collection includes both Canadian and Alaskan objects. More than one hundred Canadian Inuit artifacts and carvings acquired by the Haights on trips to the Arctic in the summers of 1953, 1961, and 1962 were given to the College in 1963. The group includes skin scrapers, knives, fishing equipment, tools, and toys, as well as stone, bone, and ivory carvings by Inuit artists from Kinngait (Cape Dorset), Povungnituk, Taloyoak (Spence Bay), and Iqaluktuuttiaq (Cambridge Bay) that date from the late 1950s to about 1960.[12] The stone sculptures represent relatively early examples of an Inuit art form that saw its beginnings in the early 1950s with encouragement from James Houston and funding from the Canadian federal government and the Canadian Handicrafts Guild.[13]

Alaskan material donated by the Haights is comprised of bone and ivory carvings, labrets, arrow straighteners, projectile points, needle cases, fish lures, and other items collected in 1967 at Sivuqaq (Gambell, St. Lawrence Island). The Haights visited Sivuqaq in the summer of 1967 with the explicit purpose of adding to their collec-tion of Inuit artifacts for Dartmouth. Unlike the artifacts they collected in the Eastern Arctic, most of which were in contemporary use at the time of their collection, the material collected by the Haights from Sivuqaq is archaeological and was recovered from Sivuqaq sites excavated by local native people.[14]

Inuit soapstone carvings from the mid-1950s collected and donated by Louise Potter of Thetford, Vermont, complement the Inuit sculptures of the Haight collection. In addition there is a soapstone carving that was purchased by the College in 1957. Together they comprise an interesting group of early examples of this newly developed Inuit art form, which began to attract serious attention from museums and scholars in the 1980s.[15]

Athabaskan material in the Hood Museum of Art comes primarily from the donations of Robert McKennan, a member of Dartmouth's Anthropology Department for more than forty years and head of the College's Northern Studies program in the 1950s, Lt. Col. Alfred Clifton, Professor and Mrs. Trevor Lloyd, Professor Elmer Harp, and Mrs. Charles Sheldon. Professor McKennan's 1930s donations of Athabaskan artifacts are the College's most extensive. They include snowshoes, models, nineteenth-century beads, clothing, bows and arrow, an iron volute-handled knife, birch-bark baskets and other containers, and sheep horn spoons. Alfred Clinton's collection of about a dozen Inuit and Athabaskan objects came to the College in 1942 and, in addition to turn-of-the-century snowshoes, beaded gloves, and a belt decorated with caribou teeth, includes a nice pair of Kutchin Athabaskan moccasins from Fort Yukon made around 1910–20.[16] Several pairs of Canadian Athabaskan gloves and moccasins made in Hay River and Fort Simpson in the Northwest Territories were donated by the Lloyds, and a number of Tutchone Athabaskan moccasins and other beaded items were purchased in the Yukon Territory in 1948 by Elmer Harp, specifically for the Dartmouth College Museum. Mrs. Charles Sheldon's gift in 1950 of Alaskan Inuit, Tlingit, and Athabaskan material collected by her husband includes more than a dozen examples of beadwork and porcupine quillwork on bags and various articles of clothing. Some of the material collected by Sheldon, including a wood bowl and a sheep horn spoon, have been classified by the museum as Eyak.[17]

Archaeological material from Canada's Coronation Gulf region is represented in the museum by the donation of more than 150 bone, stone, ivory, and antler spoons, knives, harpoon heads, gravers, picks, snow goggles, spears, awls, adzes, fishing lures, bowls, arrows, fragments, and animal remains, all gifts of Elmer Harp, who explored the Coronation Gulf area in 1955 and found interior sites with Arctic small-tool materials.[18]

Although Greenland artifacts in the Hood collection number only about seventy-five, they include fine examples of clothing and beadwork. The largest group of material is that given by Mrs. William Stickney, whose husband, as a Dartmouth College freshman, was a member of Peary's 1897 Greenland Expedition. The two dozen objects include clothing—a sealskin parka, a woman's purse, *mukluks,* mittens, fine beaded collars, and women's boots—as well as several kayak, *umiak,* and sled models. Four large skin boat models were also given to the Dartmouth College Museum by longtime geography professor David Nutt. Thomas Armstrong, Dartmouth Class of 1950, donated another Greenland kayak model, as well as boots, bow and arrows, and 1950s handicrafts, and George Murphy of the Class of 1941 brought back about two dozen contemporary objects from various parts of Greenland, including clothing, harpoons, beaded cuffs and collars, and models.

Small miscellaneous donations from Dartmouth alumni, faculty, and friends abound in the northern collections: Baffin Island items from the late 1920s were given by Robert O. Fernald of the Class of 1936; Sydney Ruggles's gift of Aleut archaeological items picked up in the Aleutians during World War II came to the museum in 1948; Sally Carrighar donated an extensive group of objects, both archaeological and contemporary, collected between 1948 and 1957 in northwest Alaska; Haida argillite carvings, along with a nineteenth-century mountain goat horn spoon and wood carvings, were given to the College by Mrs. James Foster Scott in 1957 in memory of Victor Evans; a fine pre-1900 Aleut basket collected by Carl Tuttle, a U.S. Commissioner of Alaska prior to 1900, was donated in 1968 by his nephew, Warren Prosser Smith of the Class of 1913; Don Foote, a 1953 alumnus, gave the College nearly one hundred objects, mostly archaeological, from the Point Hope area in 1963; and Corey Ford, who moved to Hanover in 1952 and became a true friend of the College and its students, left his large collection of Alaskan ivory carvings, as well as baskets, mittens, models, fishing lures, and many other objects collected on his travels in Alaska and British Columbia in the mid-1930s and the early 1940s, to Dartmouth upon his death.[19]

Dartmouth's Hood Museum of Art does contain outstanding northern objects acquired by other means, among them an exceptionally fine Northwest Coast knife collected by George Emmons and acquired by exchange with the Heye Foundation, and several late-nineteenth-century kayaks from West Greenland and Nunivak Island given to the College by the Smithsonian Institution. Other fine items, such as the Hood's turn-of-the-century Kwakiutl frontlet, cape, and tunic acquired in 1987, have been purchased over the years. Such purchases are likely to continue. As Tamara Northern has said, "Dartmouth's more recent focus on Native American art as a corollary to cultural documentation presents a new challenge to further build and extend our Native American collection."[20]

As is the case with most institutions, particularly outside their primary areas of collection, Dartmouth College for the most part has had the generosity of its friends to thank for its extensive northern holdings of indigenous art. The kind of representative, contemporary objects from the early and middle decades of this century that make up much of Dartmouth's northern collections—and that are eminently useful for teaching purposes and as cultural documentation—were readily available and affordable to Dartmouth faculty, students, alumni, and friends working in or visiting the North throughout that period, but those same objects would be both prohibitively expensive and difficult to acquire today. Prominent contemporary Canadian Inuit artists represented in the museum's collection include Kenojuak (a skin stencil entitled *Dogs See the Spirits*), Innukjuakjuk (the stonecut print *Female Owl*), Mary Ashevak (the stonecut print *Woman Sewing Skin Boots*), and Kunu (the stonecut print *Boy with Ball*). All of these works are from 1960.

## The Stefansson Collection

Many of Dartmouth's Inuit objects were either given to the Stefansson Collection or came to the College as a result of that collection, now recognized as the world's leading resource on Arctic exploration. Though virtually all of that material has now been transferred to more appropriate storage at the Hood Museum of Art, a number of paintings, drawings, and prints by Western artists remain with their relevant manuscript collections and related artifacts in the library.

Perhaps the most interesting example of northern art in the Stefansson Collection is a group of seven paintings and numerous sketchbooks by Belmore Browne, maintained with a collection of his papers, which were given to the Dartmouth College Library by his daughter in 1969. Browne was a well-known writer, explorer, and mountain climber in Alaska and Canada. He made his first trip to Alaska in 1889 and spent much of the rest of his life hunting, climbing, and painting in the North, as well as writing about it for a wide audience. The sketchbooks in the collection are particularly interesting, as they contain a great many fully developed drawings and studies of animals, landscapes, and camps in a variety of media, drawn in the field in Alaska and British Columbia in the first decade of this century. His daughter, Evelyn Browne, has extensively annotated the sketchbooks to make connections between the drawings and well-known paintings by Browne in his books and in public collections. Though much has been written about Browne, and one publication reproduces some of this material,[21] much of it has never been published or exhibited.

Other fine-art materials accompanying the manuscript collections include the five watercolors, wash drawing, and colored pencil drawing by Arctic explorer, artist, and telescope maker Russell William Porter. The watercolors of people, landscapes, boats, and icebergs in Hudson Strait and elsewhere in the Arctic were donated to the collection by Porter's friend and colleague John C. Pierce of Plainfield, Vermont, who relates that Porter said he did them with glycerin added to the water to prevent its freezing while he worked.[22]

Much of the finest imagery of the Arctic is contained in books published as official or unofficial accounts of voyages of exploration. The Stefansson Collection of books abounds in early volumes with beautiful Arctic imagery, but as discussed earlier, published books, even when they contain engravings and lithographs, are outside the scope of this study. Housed in the Stefansson Collection, however, are several groups of lithographs and other prints in portfolio, essentially without text, that do fall within the range of this search. One of the most interesting examples is the set of five lithographs in portfolio, drawn by Benjamin Russell and lithographed by John H. Bufford, entitled *Abandonment of the whalers in the Arctic Ocean Sept. 1871.* They are proof copies of five different images made from drawings by Russell, illustrating the abandonment of thirty-four New Bedford whaling ships off Point Belcher, one hundred miles south of Point Barrow, in 1871. These were the last five prints done by Russell[23] and are striking illustrations of this well-known story, the greatest disaster of the New Bedford whaling fleet and a harbinger of the eventual decline of the industry.[24]

A similar example of fine Arctic imagery in published lithography is *Ten coloured views taken during the Arctic expedition of Her Majesty's ships 'Enterprise' and 'Investigator,' under the command of Captain Sir James C. Ross,* drawn by W. H. Browne and lithographed on stone by Charles Haghe. The bound quarto of views on seven plates was issued by Ackermann of London in 1850 and is one of the most impressive examples of the "Arctic sublime" sensitivity among artists accompanying early expeditions to the Far North.

Though its book format places it at the margins of the scope of this study, it must be mentioned that the Stefansson Collection has among its rich holdings of published Arctic material a fine copy of that most important and scarce volume of early Arctic photographs, *The Arctic Regions,* by William Bradford. Published in London in 1873, the volume was based on the prominent painter's most ambitious expedition to the Arctic, in 1869. An extraordinary photographic and artistic achievement, it is of folio size and was limited to three hundred copies, each with 125 original photographs hand-tipped into the text.[25]

---

Thanks to a long tradition of interest in Canada and the Arctic, an impressive record of research and activity in northern regions, and the generosity of its alumni, faculty, and friends, Dartmouth College has a wealth of material relating to the circumpolar North. During the peak years of Arctic interest at the College, and especially during the 1950s and early 1960s, when Vilhjalmur and Evelyn Stefansson were on campus to focus that interest, the holdings of ethnographic and fine art material grew rapidly into sizable and useful collections.

Librarians of the Stefansson Collection of books and manuscripts have in the past two decades restored its vigor and its purpose, aggressively pursuing the acquisition of important materials.[26] Similarly, the opening in 1989 of a study-storage space in the Hood Museum of Art, the transfer of many items from outlying sites to this space, and the availability of online catalog access to information about museum holdings all have contributed to an increased use of the College's northern collections. With the founding of the Institute of Arctic Studies in September 1989, Dartmouth College renewed its commitment to northern activity and research.

## NOTES

1. Kesler H. Woodward, a former curator at the Alaska State Museum, is Professor of Art, Emeritus, at the University of Alaska, Fairbanks. He was director of Arts and Humanities Projects for the Institute on Canada and the United States at Dartmouth College from 1988 to 1991.

2. For an overview of this period of activity and an analysis of the growth and decline of northern research interest at Dartmouth, see Oran Young's "The Past as Prologue: A History of Arctic Studies at Dartmouth College," *Northern Notes* 1 (November 1989).

3. For more information on the growth of the Stefansson Collection and an account of its acquisition by Dartmouth College, see Evelyn Stefansson, "A Short Account of the Stefansson Collection," *Polar Notes* 1 (November 1959).

4. Vilhjalmur Stefansson died in 1962. Evelyn Stefansson Nef left Dartmouth College in 1963.

5. Only (approximately) 150 northern ethnographic items have come to the College since 1969, with two sizable donations accounting for more than one hundred of them: Alaskan ivory carving and a variety of Canadian Plains and Northwest Coast material were given by Glover Street Hastings III in 1981, and a substantial amount of primarily Plains and Great Lakes material came from Guido R. Rahr Sr. in 1985.

6. Gregory Schwarz notes in Hood Museum Accession File, 1982.

7. "Dartmouth Gets Indian Relics," *Union* (Manchester, N.H.), September 12, 1946.

8. A study of Clara Churchill's journals was done as an unpublished paper in Dartmouth's Master of Arts in Liberal Studies program by Caddie Johansen in 1987. Entitled "Images of Native Alaskans in the Churchill Collection Perceptions at the Turn of the Century," the paper provides a good overview of the Churchills' northern visit, their collecting activity, and their perceptions of Native Alaskans.

9. For more on these relatively rare baskets found at the meeting point of Northwest Coast and Inuit cultures, see Molly Lee, "Pacific Eskimo Spruce Root Baskets," *American Indian Art* (Spring 1981).

10. Tamara Northern, Curator of Ethnography, overview of the Hood ethnographic collections in *Treasures of the Hood Museum of Art* (Hanover, N.H.: Hood Museum of Art, Dartmouth College, 1985).

11. For more on the Rasmussen Collection at the Portland Art Museum, see Robert Tyler Davis, *Native Arts of the Pacific Northwest, from the Rasmussen Collection of the Portland Art Museum* (California: Stanford University Press, 1954).

12. See *Polar Notes* 6 (1966): 46.

13. A great deal has been written about the late development of this art form. Among the best references are Nelson Graburn, "Traditional Economic Institutions and the Acculturation of the Canadian Eskimos," in *Studies in Economic Anthropology,* American Anthropological Association Studies 7 (Washington, D.C.: 1971), pp. 107–121; his "Commercial Inuit Art: A Vehicle for the Economic Development of the Eskimos of Canada," *Inter-Nord* 15 (1978): 131–42; and his "The Eskimo and 'Airport Art,'" *Transaction* 4, no. 10: 28–33.

14. Letter from Sherman Haight to then-director of the Dartmouth College Museum Elmer Harp, November 27, 1967, Hood Museum accession file.

15. For an interesting discussion of contemporary Inuit art in southern Canada and elsewhere, see Peter Millard, "Contemporary Inuit Art—Past and Present," *American Review of Canadian Studies* 18, no. 1 (1987): 23–29.

16. Kate Duncan of Seattle University, who is an authority on Athabaskan material in general and beadwork in particular, reviewed photographs of Athabaskan materials in the Hood Museum of Art collection in 1985. The 1910–20 dating and several other identifications of Athabaskan materials in the collection come from a February 1985 letter to Tamara Northern, then Curator of Ethnography at the Hood Museum.

17. An interesting unpublished paper, "The Sheldon Collection," done for Dartmouth's Arctic Seminar by student R. Duncan Mathewson in 1960, with the advice of Dr. Alfred Whiting of the Dartmouth College Museum, discusses the Sheldon materials at length and speculates on their classification as "Eyak."

18. See Elmer Harp, "Prehistory in the Dismal Lake Area, N.W.T., Canada," *Arctic* 11, no. 4 (1958): 219–49.

19. For more information on Corey Ford at Dartmouth, see Mildred C. Tunis, "The Corey Ford Collection," *Dartmouth College Library Bulletin* (April 1973): 109–122.

20. Northern, overview in *Treasures of the Hood Museum of Art*.

21. Robert Hicks Bates, *Mountain Man: The Story of Belmore Browne, Hunter, Explorer, Artist, Naturalist, and Preserver of Our Northern Wilderness* (Clinton, N.J.: Amwell Press, 1988).

22. Handwritten note from the donor of the works of art, John C. Pierce of Plainfield, Vermont, accompanying the art work in the Russell William Porter Papers, Stefansson Collection Mss 118, Special Collections, Dartmouth College Library.

23. For more on Russell and these well-known prints, see Elton W. Hall, *Panoramic Views of Whaling by Benjamin Russell* (New Bedford, Mass.: Old Dartmouth Historical Society, 1981), Sketch Number 80.

24. For a good account of the disaster, see Harold Williams, ed., *One Whaling Family* (Boston: Houghton Mifflin, 1964).

25. For more on Bradford and *The Arctic Regions,* see John Wilmerding's exhibition catalogue for the DeCordova Museum and New Bedford Whaling Museum, *William Bradford, Artist of the Arctic* (Lincoln and New Bedford, Mass.: 1969).

26. For a history of the Stefansson Collection, including a discussion of current acquisition policies and cataloging projects, see Philip Cronenwett's article, "The Stefansson Collection," in *Northern Notes* 1 (November 1989).

Cat. 48. Kodiak Island, Alaska, kayak (*baidarka*) miniature with three men, two harpooners and one on a paddle, collected mid-19th century, skin, wood, paint, thread; 13.1.591.

# Thin Ice: Inuit Life and Climate Change

NICOLE STUCKENBERGER

The far north is neither barren nor empty. Abundant wildlife soars in the sky, whirls in the waters, roams the ice, and teems across the land. When the spring sun shines, seals and their offspring bask on the sea ice. In the summer, berries ripen during the long, warm days. When the autumn fogs cover the freezing lakes, streams, and sea, large herds of narwhals migrate through the ocean and countless caribou roam the tundra. During the long winter nights of icy cold, seal snouts can be seen breaking through breathing holes in the sea ice before retreating again into the Arctic waters.

The Arctic climate zone covers Greenland and the northernmost parts of the Eurasian and North American continents, including regions of Canada, Alaska, Siberia, and Scandinavia. It is characterized by long, harsh winters and short, intense summers. The poles experience strong seasonal variations in the length of day and night. The landscapes of the Arctic zone vary remarkably during the year, alternately snow covered and grass laden, blanketed in permafrost and filled with wildflowers, thick with frozen sea ice and rolling with slate-blue waves.

Cat. 46. Southwestern Greenland, miniature *umiak* with four women and one man, a dog, and camping equipment, collected prior to 1955, sealskin, wood, cloth, metal pieces, hair(?). Gift of David Nutt; 159.9.14387 (detail).

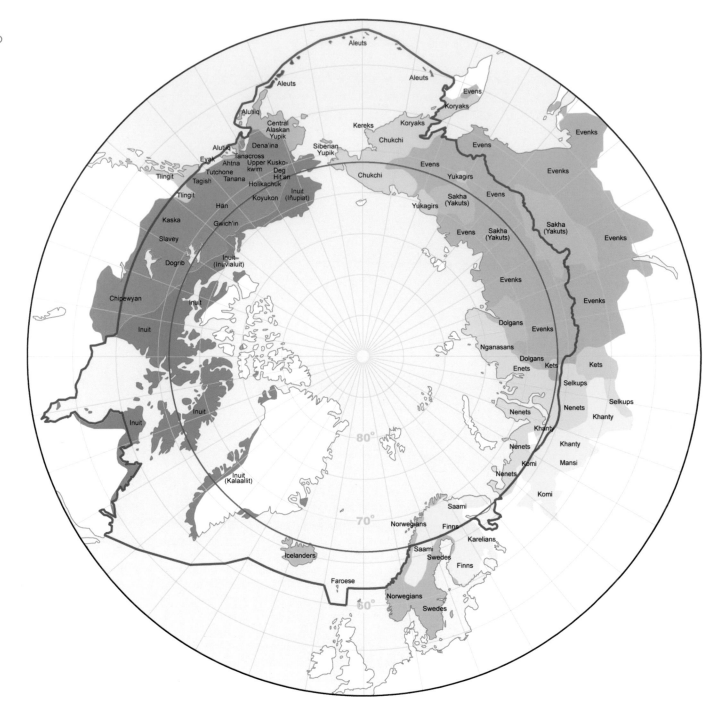

## Arctic peoples subdivided according to language families:

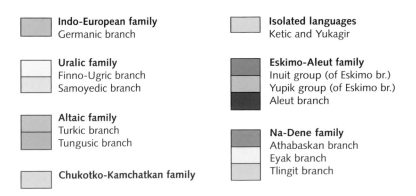

Indo-European family
Germanic branch

Uralic family
Finno-Ugric branch
Samoyedic branch

Altaic family
Turkic branch
Tungusic branch

Chukotko-Kamchatkan family

Isolated languages
Ketic and Yukagir

Eskimo-Aleut family
Inuit group (of Eskimo br.)
Yupik group (of Eskimo br.)
Aleut branch

Na-Dene family
Athabaskan branch
Eyak branch
Tlingit branch

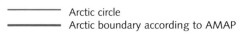

Arctic circle
Arctic boundary according to AMAP

### Notes:

Areas show colors according to the original languages of the respective indigenous peoples, even if they do not speak their languages today.

Overlapping populations are not shown. The map does not claim to show exact boundaries between the individual language groups.

Typical colonial populations, which are not traditional Arctic populations, are not shown (Danes in Greenland, Russians in the Russian Federation, non-native Americans in North America).

Fig. 3. Demography of the indigenous peoples of the Arctic, based on linguistic groups. Compiled by W. K. Dallmann, Norwegian Polar Institute, P. Schweitzer, University of Alaska, Fairbanks. Designer: Hugo Ahlenius, UNEP/GRID-Arendal (http://maps.grida.no/go/graphic/).

The Arctic is home to about four million people, both indigenous and more recently arrived from southern regions, living in towns or on the land as hunters, fishermen, herders, or, most commonly, some combination of these occupations. The Arctic indigenous peoples have distinct but sometimes related languages and cultures (fig. 3). The largest groups are the Inuit (Eskimo) peoples of Greenland, Alaska, Canada, and northeastern Siberia; the Athabascan groups in Alaska and Canada; the various Siberian peoples, such as the Yakut or smaller groups like the Nganasan and Nivkh; and the Saami (Lapps) of northern Scandinavia. Traditionally, these indigenous peoples maintained a subsistence lifestyle, and their social lives, economic practices, and spirituality were profoundly shaped by the unique Arctic seasonal cycle and its daily weather conditions. Despite numerous changes, both cultural and environmental, Arctic life remains strongly interconnected with climate (fig. 4).

The Hood Museum of Art collections contain nineteenth- and early-twentieth-century objects that reveal the Inuit people's profound involvement with their environment through the practice of hunting. The Inuit developed highly specialized hunting techniques to effectively harvest and utilize animals and fish. They were in particular known for their ingenuity with materials and the complex workings of their hunting equipment in one of the most demanding climates in the world. What makes these Hood objects so invaluable in this context is their evocative embodiment of the link between the Inuit and their natural surroundings, a link that is still relevant now. The exhibition *Thin Ice: Inuit Traditions within a Changing Environment* and this accompanying catalogue attempt to look at the Hood collection through the lens of the environmental conditions of Inuit life, the environment's importance to their culture, their contact with Western culture, and the Inuit's observations of recent climate change.

The Arctic region is politically, economically, and socially integrated with varying success into eight nation-states (the United States, Canada, Iceland, Denmark [including Greenland and the Faroe Islands], Sweden, Norway, Finland, and Russia). Arctic peoples are linked to the global market economy as citizens of countries with national and international policies on resource development, trade, biodiversity, and animal rights. However, indigenous political, economic, legal, and social practices at the local level often operate on principles different from those of the relevant nation-states. A new generation of indigenous leaders now works within institutional contexts such as the Inuit Circumpolar Conference to link community practices and issues to those of the wider world. These leaders actively pursue mutually satisfying governance strategies and means of managing change while trying to mitigate the potentially disruptive impact of national and international policy on indigenous hunting, herding, fishing, and gathering practices (Young and Einarsson 2004; Nuttall et al. 2005: 662ff).

In several studies done in the last decade and reflected in the Arctic Climate Impact Assessment (ACIA),[1] members of Inuit communities

Fig. 4. Hunters observe the animal-rich floe edge; Qikiqtarjuaq area, Nunavut (Canada), spring 2000.
Photographs by Davidee Nuqingak.

have expressed concern about thinning ice, increasingly erratic weather patterns, and the greater number of extreme weather events.

Of course, people everywhere are challenged by global climate change, which affects not only ecosystems but also our ways of life, the very core of our humanity. However, a more rapid change of climate in the Arctic has already visibly altered the land, sea, ice, fauna, and flora. Northern indigenous peoples, governments, scientists, and special interest groups are working together to understand and deal with these processes of environmental change (see cat. 2).

## Climate and Weather in Inuit Life

By examining how Inuit people connect hunting and community life to weather and their environment, we may come to a greater understanding of the significance of these latter elements within this culture and, by extension, our own. Attributing meaning to weather is not unique to the Inuit. The belief in and experience of a connection between weather and wellbeing is indeed almost universal in human relations with the natural world. Culture-specific ideas and practices with regard to this equation arrive via the knowledge and traditions handed down through generations. For example, Western European culture abounds in folk, oral, or religious traditions that predict and interpret weather, ranging from whole mythologies to simple axioms ("Red sky at night, sailor's delight; red sky at morn, sailors take warning"). On Groundhog Day, if the eponymous creature comes out and sees his shadow, folk knowledge predicts six more weeks of winter. Here in New England, the Farmer's Almanac every year predicts the strength and duration of winter based on similar observations of nature, such as the width of the band on a caterpillar's back. Lastly, regardless of whether one uses satellite forecasting or observes natural phenomena to predict the weather, there is a tendency to associate it with human events. On September 11, 2001, for instance, the bright beauty and vivid blue of the cloudless sky that morning appeared incongruous in light of the horror of the day's events.

Historically, the Inuit groups of Greenland, Canada, and Alaska developed around hunting large sea mammals, seals, fish, and caribou.[2] They lived in loosely organized communities with non-institutional leadership patterns (the Alaskan groups were generally more structured, however). Hunting had always been more than a technical, social, or economic enterprise alone, as the Inuit believed that humans, animals, and various natural forces shared spiritual qualities. Relations among these beings were subject to rules of proper conduct and various ritual practices. If these relations became strained, it was the job of the angakok, or shaman, to mediate between the physical and spiritual realms. Cosmology was thus inseparable from social life and mediated in difficult cases by this powerful cultural figure.

In traditional Inuit society, everyone was his or her own spiritual master, and people received secret knowledge (the workings of amulets, auspicious hunting songs, and so on) from spirits or elders. Angakuit (shamans) were men or women who went through an initiation that might include a near-death experience through exposure, the appearance of visions, or a long apprenticeship. They then gained helping spirits and qaumaniq ("light") with which to better penetrate the shadows that hinder ordinary sight, enabling them to see the domains of spirits and souls. Shamans could also metamorphose into animals, traveling to faraway places in a new body. Their rituals reestablished broken links between humans, spirits, and animals, and they healed afflictions. Malevolent shamans used their powers for their own good, and they were much feared.

Most Inuit groups today, which still consider themselves hunting societies, formulate their self-perceptions and views on weather and climate in similar cosmological terms, emphasizing relations among people, God, the land, and animals.[3] The Arctic weather mediates among humans and animals chiefly by providing favorable or unfavorable conditions for hunting.

The need to manage nature's active and potentially threatening force through modern technology and science is prevalent in Western contemporary culture. It can be argued that while scientific forecasting and measuring instruments are integrated into Inuit decision making—especially with regard to optimal conditions for travel and hunting trips—this is not the major knowledge base that Inuit people draw on to predict weather and ice conditions. Decisions instead depend upon the exchange of individual experience and traditional knowledge among community members. Weather and climate, taken together—what the Inuit call sila (more about this shortly)—is an influential agent that pervades all aspects of Inuit life, not simply a natural phenomenon. This essay explores the Inuit understanding of climate and weather and how it has reflected larger Inuit belief systems and social conditions as they have changed over centuries.

## The Concept of *Sila*

The closest one can get to the Western notion of climate in Inuit culture is *sila,* which connotes in turn "universe," "sky," and "weather."[4] *Sila* expresses itself in the changing seasons, which in turn shape Inuit long-term expectations of, for example, weather conditions, snow and ice quality, and periods of open and frozen sea. Inuit practices of weather forecasting are based, like Western scientific practice, on detailed, systematic observation, but also on communal knowledge and personal experience. Elder Chester Noongwook from Savoonga, Alaska, suggests the following:[5]

First thing: get out early in the morning and check the wind and the sky conditions, whether the sky is cloudy, and also whether it is cold or warm in terms of your body feeling.

In the old days, we always used to go down to the seashore every morning—to check the ice and weather conditions at the water (sea level), how the current is moving, and where is the tide.

Always talk to other people about weather and ice conditions, listen to other people's minds to see whether it is good to go out hunting.

Check for any change in wind and weather condition; we were told to watch out for weather all the time, either we are on the ice or on the shore—every hour, every minute or listen to other boats what they are saying.

Keep watching for any change in water, because of currents or clouds or waves—any sign of water change is very important.

You can never make a good forecast for tomorrow if based upon today's weather. Better go out and check it in the evening. Make a guess and check it next day: it is better to see whether it is correct or not. (Krupnik 2002: 176)

These broad similarities in observational approach, though useful for collaborations between scientists and indigenous people, should not imply that "climate" and *"sila"* are analogous concepts, however. *Sila* can also be translated as "human intelligence" and "understanding." The complex concept of *sila* thus finds its way into a large variety of composite expressions, such as the following examples from Igloolik:

*Silatuvuq:* "S/he is intelligent, has understanding/sense."

*Silaittuq/silatittuq:* "Lacking *sila*," used to describe a child or an insane person; also used for someone going on a trip even though the weather is or will be bad. In this situation, *taimaililaurtuq silaittukuluugami* ("his behavior was such that it did not show any proper thinking") is also used.

*Silarqisitsiivuq:* "Someone who waits for good weather" (and thereby proves him/herself reasonable and proper).

*Silarqisiurpuq:* "Someone who is smart enough to travel with good weather."

*Silaujualuq:* "A very intelligent or reasonable person."

*Silairrivaa/silairrittijuq:* "To seem to be something else, to trick someone."

*Silaap Inua:* "Spirit master of the universe."

*Silaat:* "Earth eggs."

*Silaluttuq:* "Bad weather."

(Wim Rasing, Frédéric Laugrand, personal communication)

In Inuit pre-Christian beliefs, *sila* was associated with a spirit master named Narssuk, a giant infant who was unpredictable in behavior and temperament, just like the Arctic weather. Significantly, Narssuk responded to human misconduct by sending violent snowstorms and prolonged spells of unfavorable weather that, if not dealt with by shamans, would keep men from hunting, thus causing starvation and death by freezing (Kangok, Boki, Shaimaiyuk 2001: 220). Weather was perceived as a moral agent manifested in a physical reality, and it connected Inuit society to the universe. In the 1920s, Najagneq, an Inuk man from Nunivak Island (Alaska), related the following to Knud Ramussen, the Danish-Inuit Arctic explorer:

Fig. 5. Polar bear skin hung to dry outside a house. St. Michael and All Angels Anglican Church can be seen in the background; Qikiqtarjuaq, Nunavut, December 1999. Photo by Nicole Stuckenberger.

[The] power that we call *sila* . . . is not to be explained in simple words. A great spirit, supporting the world and the weather and all life on earth, spirit so mighty that his utterance to mankind is not through common words, but by storm and snow and rain and the fury of the sea; all the forces of nature that men fear. But he has also another way of utterance, by sunlight, and calm of the sea, and little children innocently at play, themselves understanding nothing. Children hear a soft and gentle voice, almost like that of a woman. It comes to them in a mysterious way, but so gently that they are not afraid; they only hear that some danger threatens. And the children mention it as it were casually when they come home, and it is then the business of the shaman to take such measures as shall guard against the peril. When all is well, *sila* sends no message to mankind, but withdraws into his own endless nothingness, apart. So he remains as long as men do not abuse life, but act with reverence towards their daily food. (Rasmussen 1927: 385ff)

## Continuity and Change in Inuit Life

In the last century, Inuit traditions and lifestyles have been altered by conversions to Christianity and the (often forced) relocation of small nomadic hunting camps into centralized permanent settlements (fig. 5). Outside governments based these settlements on their own societies by providing "modern" housing, a Western education, and decidedly foreign notions of community adminis-

tration, health care, and law. Nevertheless, despite this drastic social change, weather has remained an important element of Inuit life, continuing to shape daily activities such as traveling, hunting, fishing, and gathering.

The ancestors' knowledge and ways of life are an integral part of Inuit society. Their presence among the living is underscored through the passing on of their names from one generation to the next. The importance of the ancestors' ways for today's Inuit society is expressed in the recently developed notion of the *inummariit* (the "real Inuit"), who are imagined to be very strong, skilled, and knowledgeable people who could live entirely off the land and sea. *Inuit qaujimajatuqangit* ("Inuit traditional knowledge")[6] encompasses the entire cultural underpinnings of the society, including beliefs, values, perceptions, language, social organization, and life skills (Working Group 1998: 5). The people who today come closest to the *inummariit* ideal are Inuit elders who experienced life in the nomadic hunting camps of the past (fig. 6). Younger people prize these elders' memories, knowledge, and skills and seek their advice on many issues. Even the act of listening to an elder speak is thought to be spiritually, personally, and socially healing. However, as much as many young people appreciate the older members of their community, they do not always agree with them, mainly because they now wish to conduct their lives differently than was the practice in the past. Issues that have come up include advice concerning domestic violence (elders urge women with abusive husbands to stay in their marriages, while such women are increasingly seeking separation or divorce); the custom of being silent and not voicing disagreement with an elder; and the elders' wish that certain types of music, such as hard rock, not be played at dances

Fig. 6. Mialia Audlakiak and a girl; Qikiqtarjuaq, Nunavut, 2000. Photo by Davidee Nuqingak.

or on the radio. Many elders feel that some Western contemporary music is dangerous to the wellbeing of young people (Stuckenberger 2005). Many Inuit people feel such differences have created a widening gap between old and young in a society where Western standards of social conduct have had a progressively greater impact.

Regardless, "tradition" in the Inuit context is best viewed as "a dynamic concept referring to ideas, practices, and institutions that are handed down from one generation to the next and change in the process" (Oosten and Remie 1999: 2). Thus, even behavior and customs that resulted from contact with Western societies, such as drinking tea, dancing jigs,[7] or, on a much different cultural scale, practicing Christianity, have become part of the *inummariit* image, provided only that such behaviors are perceived to be both useful to and consonant with *Inuit qaujimajatuqangit.* For the Inuit it is not so much *what* a person does—people differ—but *how* he or she does it that is important (Omura 2002: 107). This is vital to keep in mind in any examination of continuity and change in the Inuit perceptions of nature and weather.

## Inuit Cosmology: *Sila* and Creation Myths

Studying Inuit mythology provides valuable insight into their traditional culture and relationship to the environment. As scholar Hugh Brody points out, "The past—even the remote past—enters the present, becomes part of it in stories, in myths . . . When Inuit of today tell their stories, talk about the past and about the first occupants of the Arctic, they are also talking about themselves" (Brody 1976: 186). In mythological stories of creation, the universe and Inuit society become existentially joined, but it is a fragile relationship. Inuit creation myths tell how the world evolved via metamorphosis and diversification; how it is ordered; and how human society is part of it. They also provide relevant background about the elemental involvement of humans with *sila.*

For the Inuit, cosmic order was born out of the turmoil of the lives of the first two human beings, men named Uummaarniittuq and Aakulukjjuusi, who entered the world when there was no daytime, death, sun or moon, seasons, light, sea ice, weather, thunder and lightning, or storms. The ground was made of earth and cov-

ered with snow (Saladin d'Anglure 1990: 82), and the life-giving *nuna* (earth) and *sila* (in this case, sky) were yet to diversify into the myriad shapes they would take as the world evolved. Thus, Uummaarniittuq and Aakulukjjuusi emerged from simple hummocks of soil and were filled with *sila's* breath (see cat. 21).[8] They had the power to procreate, and as one got pregnant, the other formed his breath into words that transformed his companion into a woman who could give birth. From the joining of these two men, just like from the joining of earth and sky, offspring came forth. Through their misdeeds, suffering, and magic words, these people then caused, among other things, death and ancestors; thunder and lightning; the sun and the moon, and thus day and night and the seasons; the weather; other celestial bodies; animals; and spirits. All of these new beings shared the world with a few species of animals that also came from the earth, or from other worlds. With the creation of each new kind of being, humans had to observe new rules, such as the prohibition of incest, the foundation of matrimonial exchange, and the obligation to take care of widows and orphans and share game. By observing these things, they maintained proper relationships between their society and these various new elements (Saladin d'Anglure 1990: 89).

With the creation of the seasons in particular, Inuit life developed around these new rhythms and variations in nature. An Inuit story tells of an incestuous relationship between a brother and sister. After their coupling in the dark of night, the sister discovers she has lain with her brother and starts running from him. During their chase, they become transformed into the sun and the moon circling the sky. This metamorphosis influenced the physical shape of *sila* as the universe, which transformed as well into the circling movements of day and night and the seasonal cycle through Brother-Moon chasing Sister-Sun.[9]

A Netsilik Inuit woman named Nalungiaq from the central Canadian Arctic told Knud Rasmussen in the 1920s of another tale in the creation myth cycle:

How came the winds and the rain, snow and storm [into being]? Once there lived giants on the earth. There were not many, but we have heard of them. These giants warred against one another, and once it happened that a giant infant became an orphan because its father and mother were killed. The boy's name was Narssuk, "Little Belly." The infant "Little Belly" was brutally snatched by malevolent humans from the warmth of his mother's parka. Instead of nurturing him, they ridiculed the giant baby. [Their] wickedness was to make him helpless, but, instead of perishing, he went up to the sky and became Sila, the weather, who took revenge on the evil people by means of gales, torrents of rain, and snow. (Rasmussen 1931: 210)

Narssuk, now the spirit master of *sila*, would become enraged about human misbehavior, particularly breaches of social rules and acts that offended animals. As a result of this myth, weather came to be viewed as unpredictable and similar to an infant's tantrums. (During prolonged spells of bad weather or extreme weather events, the Inuit would call on the shaman to appease Narssuk [see cat. 22].) In other myths, humans also metamorphose into the spirits of thunder and lightning, fog, and different kinds of wind and snow, all of which were under Narssuk's control.

As related in the myth of Sedna, the establishment of the special relationship between the Inuit people and sea mammals marks the final stage of their development as a hunting culture thriving near the coast. Sedna was a girl who did not want to marry. She was betrothed first to a dog and then to a stormy petrel in the shape of a man, by whom she was held captive on a small island. When her father came to rescue her in a boat, the enraged sea bird beat his wings, causing a storm and high waves that threatened to capsize them (see cat. 8). The father, to save himself, threw the girl overboard. As she clung to the boat, he chopped off her fingers, and an assortment of sea mammals sprung from them. Sedna then sank to the bottom of the sea, where she became the goddess of sea game, releasing the animals for the Inuit to hunt. The Inuit soon came to rely on this game and had to observe many rules and taboos to maintain good relations with their important prey (see cats. 34 and 65). Animals, though distinct from humans, share their spiritual essence and are both aware and intelligent (see cat. 24). Animals give their bodies for the procreation and perpetuation of humanity, and their immortal souls can reincarnate after death, provided that they are treated with respect. Anthropologist Bernard Saladin d'Anglure notes, "From [Sedna's transformation] on the animals are food, and humans are spouses, parents, allies or enemies. The mythical realities still exist. . . . [But they are] most easily accessible by shamans

and spirits" (Saladin d'Anglure 1990: 93). Thus, hunting guarantees the renewal of Inuit society in its physical, social, and cosmological aspects, and it is a key to understanding Inuit spirituality as well.

In the world that existed after Sedna became the goddess of sea mammals, spirits continued to interact with humans. Narssuk, in addition to being the spirit who expressed himself through weather, was also believed to complete the creation of human life by inserting the *tarniq* ("soul") into an unborn baby. In Igloolik the *tarniq* was visualized as a miniature image of the person, existing in an air bubble that was filled with life-giving *sila*. Narssuk inserted the bubble into the child's liver and would regain it again at death (Saladin d'Anglure 1980: 37).[10] Narssuk therefore exists both outside people, by causing weather conditions (and, if he is angry, interfering with the procurement of food through hunting), and within them, as a component of the soul. This demonstrates the comprehensive reach of *sila* in Inuit culture. In fact, the seasons, the weather, and human society are all interconnected expressions of a mythological, moral, and physical universe that evolved in Inuit myth as *sila* diversified through the metamorphosis of the first human beings and animals into stars and planets and their movements; through animals and their ways; and through the weather and its infantile character.

The term *sila* thus encapsulates a multitude of meanings, some even seemingly at odds. How can a single concept express something as vast as the universe and as tiny as the bubble that holds the soul? Saladin d'Anglure answers this question:

This apparent contradiction has to do with the fact that *sila* is a relative term and must be analyzed within the specific contexts of its use: on the level of the minuscule, because *sila* provides that portion of air that surrounds the human life-soul *tarniq*. At death, the bubble bursts and the life-soul travels to another world of eternal life; the liberated air becomes part of the atmosphere of the cosmic *sila*. For the *tarniq, sila,* therefore, puts the "outside" into a bubble and into the living body. On the level of human life, *sila* is also the immediate outside touching the body and the surrounding air. On the cosmic level, *sila* is the great exterior under the celestial cupola. (Saladin d'Anglure 1986: 71, translation by author)

*Sila* thus animates existence at every level, diversifying creation while also binding it together. People grow into a state of *silatujuq* ("gifted with *sila*," or, more generally, "understanding") through interacting with others and exposing themselves to *sila* by living off the land (Saladin d'Anglure 1980: 37; Oosten 1989; see cat. 44).[11]

As scholar Britt Kramvig observed for the Saami of Scandinavia, nature is not the "other" to culture and humanity for these peoples (Kramvik 2003: 168). In the Inuit's holistic society (both past and present) all elements of life—relationships to the land, to animals, to community members and ancestors, and to their economic, aesthetic,[12] and cosmological underpinnings—are interconnected and cannot be viewed in isolation. However, today's Inuit communities provide very different social and living conditions than did the nomadic hunting camps. Settlements now provide food and housing (through both a market economy and government subsidy), and Inuit people can live without going out on the land at all. However, many still prefer the *inummariit* ways and seek food (and identity) via a more traditional lifestyle (see cat. 45).

## Conversion to Christianity: A Case of Continuity and Change

The Inuit conversion to Christianity occurred in different places at various times from the eighteenth through the early twentieth centuries. In southern Baffin Island, for example, the Reverend James Peck established the first Anglican mission in 1894 at Uumanarjuaq (Blacklead Island in the Cumberland Sound), a region famous for playing host to European and American whaling efforts. For years Peck had little success among the Inuit (Oosten, Laugrand, and Kakkik 2003: 32), whose elders and camp leaders were not amenable to the new teachings (Laugrand 2002b: 117). Angmarlik, a respected hunter, prominent camp leader, shaman, and famous whaling captain with the fleets, fiercely resisted Christianity until 1902, when he is said to have received a revelation from Sedna, the mistress of sea mammals, telling him to adopt the new beliefs (see cat. 33). He then became a preacher, acknowledging Sedna and adopting the Christian "rules," which he believed to be powerful agents in dealing with social and cosmological relationships. Many of the old practices and beliefs about the universe, animals, and spirits were replaced, and most of the Inuit rules pertaining to hunting, birth, and death disappeared, along with their shamanic practices. Sedna and the other spirit beings who withheld sea game when people transgressed social and ritual rules were then replaced by the Christian God, who had created animals for the benefit of human beings and confirmed hunting as the true vocation of the Inuit way of life (Laugrand 2002b: 129ff).

## Case Study: The Inuit Community of Qikiqtarjuaq, Nunavut

Students at Nunavut Arctic College asked an elder named Nutaraq from Mittimatalik (Pond Inlet) if he thought changes were happening more quickly nowadays: "Yes. There has been a lot of change happening over the years. You notice that there are always new things happening, that changes are always occurring. People change as well." (Kappianaq and Nutaraq 2001: 154)

Starting in the 1960s, new Inuit settlements not only drastically changed social conditions but also affected people's interactions with the environment. They now spend less time out on the land, due to employment and mandatory school attendance as well as the costs of hunting with equipment such as snowmobiles. Settlements are often located far from hunting grounds. Many families can no longer afford to go hunting but often support a hunting relative to provide them with "country food" (fig. 7).

Fig. 7. Hunter returns from hunting by snowmobile and *qamutik* ("sled") toward the community of Qikiqtarjuaq, Nunavut, spring 2000. Photo by Nicole Stuckenberger.

Fig. 8. Community feast during the annual Fishing Derby; Nalusiaq Lake, Qikiqtarjuaq area, Nunavut, May 2000.
Photo by Nicole Stuckenberger.

Reacting to the development of economic and social problems like unemployment, substance abuse, and suicide, the Inuit became politically organized during the 1960s. In Canada these efforts eventually resulted in the 1999 establishment of the new Canadian territory of Nunavut ("Our Land"), providing the Inuit with self-administration.[13] The territory's name expresses that the land remains vital to their culture. Inuit relations to the land and the animals, and the whole notion of *inuit qaujimanituqangit,* define their identity and are crucial to the governance of, and improvement of social conditions in, Inuit communities today (fig. 8).

The people of Qikiqtarjuaq, Nunavut, often contrast their life in today's settlements with being "out on the land." Whereas they view life in hunting camps with its *inummariit* ways as beneficial for the individual and community, many Inuit people experience the settlement ambivalently; though they enjoy being together, they also perceive the sedentary living conditions as potentially detrimental to their wellbeing.

When I interviewed Billy Arnaquq, a hunter and priest in his late forties, and Lootie Toomasie, the community's mayor in 2000, about the most important aspects of life in Qikiqtarjuaq, both immediately referred to the seasons (see cat. 5). If I wanted to learn about Inuit ways, I should stay for at least a year to experience all of the variations of *sila,* as life would differ profoundly over that time period as the community moved between the settlement, where all lived together, and the land, where people dispersed to camps with their families. For Arnaquq the seasons were marked by a combination of climatic conditions, the ice and ocean, animal migrations, and people's reactions to them all:

Elders have a passion for spring, because the winter is so long and so cold. People always have a passion for when the spring is coming, the birds are coming, the animals are more plentiful. It's a time to go out and be part of, what you may say, the celebration. That is also a time that many people get together. [Hunting in that season] is something that is just so much part of you that you just go try to do it. . . . It is like that season comes for you to enjoy it. It is inside the people. Some men have a passion for early fall, when the harvest is plentiful and the animals are migrating through. Some have a passion to go out seal hunting in the winter. Some of them cannot wait until the ocean is frozen. . . . Once you are part of that, it just becomes part of you.

The Inuit sense of the seasons is now coupled with a modern Christian calendar identifying days of rest, festival, and work. (For information about pre-settlement seasonal variations, see the introduction to the annotated checklist in this catalogue.) Together they provide the rhythm of modern Inuit society (see cat. 35). What follows is a description of some of the typical seasonal activities that I observed when staying in Qikiqtarjuaq in 2000.

## *Upirnaaq* (Spring)

As the sun rises above the horizon after the long, dark winter months, it becomes warmer. The snowy surface of the land and sea ice melts during the day and freezes over at night, forming a hard crust that provides optimal travel conditions during the nighttime and early morning hours. In late May the sea ice starts to break up. Inuit people continue traveling by snowmobile until the puddles on the ice become too deep, the ice itself

Fig. 9. Spring camp; north of Qikiqtarjuaq, Nunavut, 2000.
Photo by Davidee Nuqingak.

too thin, and the cracks too broad. They then prepare for the summer boating season.

After a more or less sedentary winter, people are happy to start camping in early and especially mid-spring (fig. 9). Most families have camps; those who work in the settlements stay for at least a few days, but others stay for several weeks at a time. Some may even arrange for babysitters for their school-aged children during their absence.

Fig. 10. Children play tag with the waves. Qikiqtarjuaq, Nunavut, 2000. Photo by Davidee Nuqingak.

## Aujaq (Summer)

The snow-free, ice-free season of *aujaq* lasts from July until September. In early summer, the muddy soil quickly dries and the landscape turns from white and grey to brown and green, with colorful patches of flowers. The sound of rushing rivers breaks the silence of the Arctic winter and early spring. The last ice floes drifting on the water melt rapidly. People transport boats from settlements to the shore and embark on camping, net-fishing, and berry-picking trips. Boating in early summer is not without peril, as collisions with drifting ice can easily damage propellers. It requires skill and experience, as well as at least two people, to maneuver a boat through ice floes (see cat. 46). Lootie Toomasie explained, "During the summer, we are more scattered on the land, because the weather is warm compared to [winter]. We have twenty-four hours of daylight, sunshine in June and July, for us to enjoy being away from the community during that time. Being away from the community is really good for those of us who keep continuing the traditional ways of life . . . If we have to go out on the land, nobody can stop us!"

Arctic white heather (*Cassiope tetragona*) added to a campfire produces a flavorful smoke that is absorbed by brewing tea, called *igaaq*. This smoked tea is served with bannock (bread baked in a pan) and jam, a much-anticipated seasonal delicacy. Women who remain behind for work or other reasons organize tea parties around the settlement to enjoy the flavor of summer through *igaaq*. Edible seaweed such as devil's apron (a species of *Laminaria*) washes ashore before the freeze-up of autumn. On windy days, women walk along the shoreline to collect it (fig. 10). Summer is also the time of the annual deliveries by sea of gasoline and oil, prefabricated houses, cars, snow scooters, bulk wares, private food orders, and so on.

## Ukiaq (Autumn)

The first snow falls on the mountains in early September, and mist hovers over the sea. The sky is cloudy, and it is windy. Some men go caribou hunting in a region about two days' journey to the north. By the end of September, most families are back in the settlements. Sealskins dry on frames leaning against the houses, and women start to prepare winter garments. Evening dances, which were attended by only a handful of people earlier in the summer, are now busy. Church services take place more regularly and, except for the period of the communal narwhal hunt in mid-October, are also well attended (fig. 11).

In late October, heavy snowstorms sweep the open sea. The community breakwater starts to freeze up, and boulders on the beach are covered by layers of ice. The last leaves of seaweed have washed ashore, but hardly anyone picks them up anymore. Boats are taken from the water. For some six to eight weeks, traveling by sea or land

Fig. 11. Narwhals caught by hunters are about to be landed; Qikiqtarjuaq, Nunavut, 2000. Photo by Nicole Stuckenberger.

is very difficult or impossible. Most Qikiqtarjuar-miut remain in the settlement during this period. Conflicts seem to grow fiercer than during other times of the year, and some people told me of the horror of friends or relatives committing suicide.[14] This is not new: even in the late nineteenth century Franz Boas (1888) noted that late fall and early winter constituted a fragile phase in Inuit community life, when the spirits of the recently deceased threatened the community and Sedna grew weary of human misconduct and contemplated not sending any more animals from her abode in the sea.

### Ukiuk (Winter)

From around mid-December the sea is frozen solid and the wind ceaselessly sweeps the snow off the ridges, rearranging the drifts. The midwinter sun does not rise above the horizon. "Highways" form on the snow-covered land and sea ice where people use the same scooter tracks as they travel (fig. 12). Few people travel for long periods, and only a few hunters go out to catch seals at their breathing holes at this time of year because of the subzero temperatures. Many instead choose the comforts of their warm houses and the supermarket. Winter brings the major community festival of Christmas, Quviasuvik ("a time of happiness"), an occasion of spiritual and social renewal (fig. 13; Stuckenberger 2005).

Fig. 12. Traveling in wintertime; south of Qikiqtarjuaq, Nunavut, 2001. Photo by Nicole Stuckenberger.

The modern community of Qikiqtarjuaq, although sedentary in its settlement structure, continues to function in many ways like a nomadic community.

Fig. 13. All have fun with a Christmas game at the Arctic College facility, Qikiqtarjuaq, Nunavut, December 2000. Photo by Nicole Stuckenberger.

The Inuit people still highly value the seasonal variations that shape their way of life. Practices of living together in the settlement in winter and dispersing in family groups over the land in spring and summer continue to be an integral part of the organization of Inuit society. The relationship to land and animals, even within a Christian framework, is still a necessary precondition for people's wellbeing and spiritual health. It is therefore beneficial to maintain a nomadic lifestyle that connects the land and the community. For many individuals, following the seasons is part of their Inuit identity. The Arctic seasons and weather patterns shaped by the dynamics of sila or, in a Western reading, by the Arctic climate continue to be integrated into Inuit ways of life and self-awareness, so that nature is never the "other" but rather integral to culture.

### Inuit Perceptions of Change in Weather Patterns

Whenever people took me out on the land, we had to be ready to leave at any moment, though we might then proceed to wait a day or two; it all depended on the ice and weather conditions, both observed and predicted. Forecasting is a challenge, and the Inuit rely in turn upon their knowledge of weather patterns, the color of the ocean water and waves, their sense of the temperature, the mist, sky, and clouds, the wind direction, the behavior of animals, and the flight of seabirds. Elders are also consulted in acknowledgment of their lifetimes of experience with sila.

The Inuit people value knowledge that is personal and specifically situated in time and space. Each person connects to others and to nature in his or her own way, and *sila* (here "intelligence, understanding") develops uniquely (fig. 14). When coming of age, each person should have enough insight and skill to live off the land, but always with proper social behavior, displaying self-control, cooperation, generosity, and modesty. A *silarqisiurpuq* is "someone who is smart enough to travel with good weather"; this person can now live off the land and grasps the dynamics of nature.

Fig. 15. The houses and school of Qikiqtarjuaq in spring 2000. Photo by Nicole Stuckenberger.

Fig. 14. A boy plays on ice sheets while learning about their stability; Qikiqtarjuaq, Nunavut, late spring 2000. Photo by Davidee Nuqingak.

In today's Inuit society, children spend more time acquiring Western knowledge and ways of thinking at school and less time out on the land. Elders often worry that younger generations will be less and less like "true Inuit." Community officials, parents, elders, and teachers try hard to forestall this by developing school (fig. 15), college, and church youth programs for transmitting land-based knowledge and skills (traveling, hunting, sewing, camp keeping, and so on), both by actually practicing them and by telling stories about them (see cat. 43). The continuation of the Inuit hunting life is seen as crucial for dealing with the serious social difficulties faced by communities as well as for increasing individual wellbeing.

Yet the environment that once sustained that life may soon no longer exist, as the title of a leading book on traditional ecological knowledge points out: *The Earth Is Moving Faster Now: Indigenous Observations of Environmental Change* (Krupnik and Jolly 2002). Since the 1980s, Inuit communities from most of the Arctic regions have been reporting increasingly unfamiliar sea ice conditions and weather patterns, as well as more extreme weather events. In the past, such signs of Narssuk's displeasure were dealt

with by shamans and hopefully passed quickly. Today, these features seem to have become endemic. In recent years, research ethics have changed, as scientists and Inuit together have made a policy of involving the local population in the planning, organization, and execution phases of research in the Arctic. There is an increasing awareness among Western scientists of the value of local knowledge and observation for gaining insight into environmental and climatic processes and developments. In 1995, for example, geographer Shari Fox started to collaborate with the Nunavut communities of Iqaluit and Igloolik to document and communicate Inuit observations and perspectives of climate and environmental change. Five years later, the communities of Kangiqtugaapik (Clyde River) and Qamani'tuaq (Baker Lake) joined the project. Inuit observations of changed weather patterns are often framed in personal narratives of family history and practices of living off the land. This is evident in Fox's interview with elder L. Nutaraluk from Iqaluit in 2000 on the changes that he observes. Nutaraluk relates the following:

The weather when I was young and vulnerable to the weather, according to my parents, was more predictable, in that we were able to tell where the wind was going to be that day by looking at the cloud formations . . . Now, in the 1990s and prior to that, the weather patterns seem to have changed a great deal. Contrary to our beliefs and ability to predict [by] looking at the sky, especially the cloud formations, looking at the stars, everything seems to be contrary to our training from the hunting days with our fathers. The winds could pick up pretty fast now. Very unpredictable.

[The winds] could change directions, from south [to] southeast in no time. Whereas before, before the 1960s when I was growing up to be a hunter, we were able to predict. (Huntington and Fox 2005: 82)

In Nunavut and elsewhere the Inuit people have decided to take an active role in the documentation of their traditional knowledge. For instance, the Nunavut Arctic College in Iqaluit has developed the Inuit Language and Culture Program, inviting anthropologists to cooperate with Inuit students in teaching and developing interview and research techniques.[15] The Oral Traditions Course from January 2000 focused on traveling and surviving on the land, and one student asked elder Cornelius Nutaraq from Mittimatalik: "Have you noticed any changes in the weather from the time you were young up to today?" Like Nutaraluk, Nutaraq carefully considered his answer, framing his observations according to Inuit standards of reliable knowledge, which is based on knowing who noticed something as well as what they noticed and how, when, and where they noticed it:

I was more aware of the weather back then because I looked out more often. I feel the weather is different now from back then. I suspect that when I was a young child, I thought that the days were longer, there was no wind blowing, and the sun was always shining. Perhaps that is the perception I had, because I was a child. Perhaps it is because I am more aware of boats now, I want it to be calm, that I see the weather differently. The wind seems to blow more frequently and the duration of calm water seems to be shorter. Because I'm more aware, the direction of the wind seems to shift more frequently, also. Maybe this is not so, but that is how I perceive it. I have also heard this from other people, older than I am. They, too, have commented that the wind seems to blow more frequently. (Kappianaq and Nutaraq 2001: 152)

In 1997, scholar Natasha Thorp from the Tuktu and Nogak Project out of Victoria, British Columbia, interviewed Frank Analok from the Kitikmeot region of western Nunavut about his observations of a changing climate. He observed, "It used to get real nice outside when I was younger, right after a storm. Right now when the weather gets bad, it seems consistent" (Thorpe et al. 2002: 226).

Due to these changing weather patterns, even place names that evolved from Inuit interactions with the land have become deceiving. Uusaqqak Qujaukitsoq from Qaanaaq, North Greenland, observes the following:

Sea-ice conditions have changed over the last five to six years. The ice is generally thinner and is slower to form off the smaller forelands. The appearance of aakkarneq (ice thinned by sea currents) happens earlier in the year than normal. Also, sea ice, which previously broke up gradually from the floe-edge towards land, now breaks off all at once. Glaciers are very notably receding, and the place names are no longer consistent with the appearance of the land. For example, Sermiarsussuaq ("the smaller large glacier"), which previously stretched out to the sea, no longer exists. (Huntington and Fox 2005: 84)

Inuit people expect the increasing instability of the weather to have serious practical, economic, and social consequences for subsistence ways of life, especially if the pace of change becomes more rapid. Hunting on thinning ice will be increasingly more dangerous, and techniques and technology may have to be altered. Because Inuit social hierarchies and relationships, gender roles, and cultural self-perceptions are closely linked to hunting and its associated practices, new social issues may arise, with additional stress placed on individuals and communities. Elders, the closest representatives of the inummariit ways of life, increasingly feel that traditional knowledge is beginning to be out of step with sila (here, "the universe"). They fear that if things change too quickly for them to keep up with, they may no longer be able to teach skills for reliable weather prediction to the community. However, people are also optimistic that they will find ways to adapt, as have generations before them.

As already discussed, in the pre-settlement past Inuit people interpreted extreme weather events or prolonged spells of unfavorable weather as arising from human misconduct; Narssuk and the other spirits inflicted retribution through the weather. Inuit today speak of sila (as "weather") when they discuss their observations with scientists, but they no longer refer to Narssuk. However, there are Inuit people who suspect that the weather might still react to human conduct. Elder Nutaraq noted some changes for the Nunavut Arctic College students:

Another thing I have noticed is that the places where people go hunting have shacks now. It seemed that because the snow is not being used as much, there is not as much of it. It's probably not that way, but that's how I think of it. Nowadays, it seems that men who want to own things have shacks built at locations where they hunt frequently. Perhaps this is because they don't want to take the time to build an igloo. Maybe it is because they want to have something to enter as soon as they get there. Another change I have noticed is that everybody seems to be in a hurry. . . . The way of life today is different from what it was then, from the times when I became aware. A lot of years have passed. . . . After twenty or twenty-five years have passed, you notice the change. (Kappianaq and Nutaraq 2001: 153)

Nutaraq associates the disappearance of igloo snow with changes in lifestyle that make Inuit people rely on goods brought in from the south and thus become unreasonably impatient (*silaittuq*, "lacking *sila*"). This leads to a life dictated by the clock rather than by the seasons, the weather, and living off the land.

Today, the Arctic climate is changing rapidly and the weather seems less predictable. The Inuit peoples (and many others) recognize the direct effects of humans on climate from the release of greenhouse gases. There is an understanding among the Inuit that recent climate change has been caused by Western lifestyles and emissions from the factories and cars so prevalent in southern cultures. They pursue policies to control these emissions to mitigate further change. But is a scientific approach the only relevant framework for understanding climate change? In the past, the Inuit believed that human conduct influenced the weather—itself a moral agent—and that the weather had a strong impact on human wellbeing. Today, the Inuit integrate modern elements into their ways of life while keeping hunting, land, animals, and seasonality central to their cultural identities. This decision to maintain a traditional lifestyle suggests that *sila* still functions as an important agent in Inuit life. As Aqqaluk Lynge stated earlier in the catalogue, there are opportunities for Western and Inuit societies to work together on the challenge presented by climate change, but only if each is sensitive to the different world views brought to the dialogue.

1. In 2004 the Arctic Council and the International Arctic Science Committee (IASC) compiled this report "to evaluate and synthesize knowledge on climate variability, climate change, and increased ultraviolet radiation and their consequences." The results of the assessment were released at the ACIA International Scientific Symposium held in Reykjavik, Iceland, in November 2004. The assessment is also available on the Web (www.acia.uaf.edu) and was published by Cambridge University Press in 2004. The Arctic Council "is a high-level intergovernmental forum that provides a mechanism to address the common concerns and challenges faced by the Arctic governments and the people of the Arctic."

2. The Inuit groups have many fundamental cultural features in common but are also distinguished by certain particularities. For the purposes of this overview, however, they will be discussed in collective terms.

3. Most Inuit people are now Christian, hence the reference here to God. The nature of the religious dimension of contemporary Inuit identity will be discussed later in this essay.

4. The Inuit concept of *sila* is complex and varies among groups. This essay focuses on the Canadian Arctic. For descriptions of similar beliefs pertaining to *sila* and weather among other groups, see Bogoras 1904 for the Chukchi of Siberia; Garber 1940 for the story of a shamanic treatment of a blizzard among the Bering Strait Inuit; Lantis 1947 for Inupiaq speakers and also the mythology and beliefs about *sila* among the Aleut; Lantis 1959 for the Yup'ik-speaking peoples in the Lower Kuskokwim and Nunivak-Nelson Island areas; and Birket-Smith 1924 for the Greenland groups.

5. Chester Noongwook (Tapghaghmii) was born in 1933 in a hunting camp close to Savoonga, an island off the Alaskan coast. When he was quite young he accompanied his father, grandfather, and brothers as they hunted on the sea ice and also by boat. He served in the U.S. Army and the Alaskan National Guard, and between the early 1950s and 1962 he was a dogsled mail carrier. He is still an active whaling captain and a member of the Alaskan Eskimo Whaling Commission (Oozeva et al. 2004: 8).

6. This notion, which emphasizes that Inuit ways should be integrated into all levels of Inuit life, including their institutions, made its first appearance with the creation of the Inuit territory of Nunavut in 1999 (Laugrand 2002a: 94).

7. A dancing music style introduced by Scottish whalers in the nineteenth century and today considered to be part of Inuit culture.

8. According to stories that come from the communities in the Northern Baffin Island area of Nunavut, earth and sky continue to bring forth special creatures called *silaaq* ("children of *sila*"), gigantic caribou whose males are grey-brown and females are white (Saladin d'Anglure 1986: 72). According to these stories, these animals hatch from eggs that come from the earth and can be found on the tundra. Elder Nutaraq from Mittimatalik (Pond Inlet) told students from Nunavut Arctic College that the *silaaq* continue to be born today: "I never have killed a *silaaq*, but I have heard that killing it can cause bad weather because the earth is grieving. I know that this seems odd, but it is said that the weather will be bad if they are broken. There will be a lot of rain, and there might be a lot of wind close to the time when the egg is broken, hindering hunting" (Kappianaq and Nutaraq 2001: 155ff).

9. The different seasonal moons provide the names of the months in the Inuktitut language. For example, *Tirigluit*, in Igloolik, denotes the month of the bearded seal pups (May/June). This spring month is marked by a sun that hangs continuously above the horizon; the birth of bearded seal pups; snow that is too soft for building igloos; the arrival of migratory birds, such as snowy owls, buntings, and snow geese; and, in former times, the movement from winter dwellings to tents (MacDonald 1998: 196). As the Inuit peoples moved seasonally to follow the animals, it was useful to know as much as possible about the coming season. In the community of Igloolik (Northern Baffin Island, Nunavut), the movements of sun and moon are still used for long-term weather forecasts. Systematic observation in light of the relationships extending from mythological times to the present informs this process. Hunter Michael Kupaq explained that the Inuit observe the movements of Brother-Moon and Sister-Sun as they compete with each other to be the first to reappear in the January sky. The sun at Igloolik is visible on the southern horizon for a few minutes as early as January 13. If at this time the moon is growing, then Brother-Moon wins, which predicts a cold spring and summer. If it is waning or absent, then Sister-Sun wins, which predicts a warm spring and summer and an early break-up of the ice (Saladin d'Anglure 1990: 103).

10. In the past, the Inuit made use of the *sila* that was contained within them, as in the following Igloolik and Arctic Quebec practice: "Everyone born on a day of good weather had the capacity to calm storms by exposing him/herself nude to *sila* and rolling on the ground with arms raised above the head while screaming: 'But where after all is my beautiful weather? *Ungaa!* [cry of an infant]'" (Saladin d'Anglure 1986: 72, transl. by author).

11. Shamans sought to acquire some of *sila's* power when trying to reach beyond the physical realities of life. The Caribou Inuit shaman from the central Canadian Arctic, Igjugaarjuk, explained to Rasmussen that the novice who wanted to enhance his clairvoyance "receives his special powers by 'exhibiting' himself to [S]ila—by letting [S]ila see him and take notice of him. One says . . . [S]ila must keep her eyes on you" (Rasmussen 1930: 51). Shamans also used embodied *sila* for healing. A common story relates how the first shaman healed by breaking wind on the affected part of his client's body. However, the emission of air was also dangerous, as the air encapsulated in the soul-bubble could escape when one was sneezing or passing gas. Then the person had to shout "*qaaq* [burst]!" to prevent the soul from leaving (Saladin d'Anglure 1986: 71ff).

12. For instance, hunting weapons were sometimes decorated to please the animals they were destined to kill. Also, the naturalistic depiction of animals and humans in today's Inuit prints and sculptures is highly valued as an expression of the artist's close connection to his or her environment.

13. Similar land claim settlements and regional governments have arisen in other Inuit groups: for example, the Alaska Native Claims Settlement Act, the Greenland Home Rule, the James Bay and Northern Quebec Agreement, and the Inuvialuit Final Agreement. Co-management arrangements were also established: for example, the Alaska Eskimo Whaling Commission, the Inuvialuit Game Council, and the North Atlantic Marine Mammal Commission.

14. Many studies have been done on the Inuit suicide rate from socio-psychological and Western clinical perspectives. However, a comprehensive study remains to be done on this complex subject.

15. The students interviewed elders on various issues (see http://www.nac.nu.ca/library/publications.htm for the various publication series resulting from these projects). See Laugrand 2002a for a historical and anthropological analysis of traditional knowledge programs as components of identity construction and indigenous historiography.

PLATES

46

Cat. 8.
Greenland
**Miniature kayak with paddler and equipment**
About 1920
Hide, wood, fabric, bone
Purchased through the Hood Museum of Art
Acquisitions Fund; 2005.56

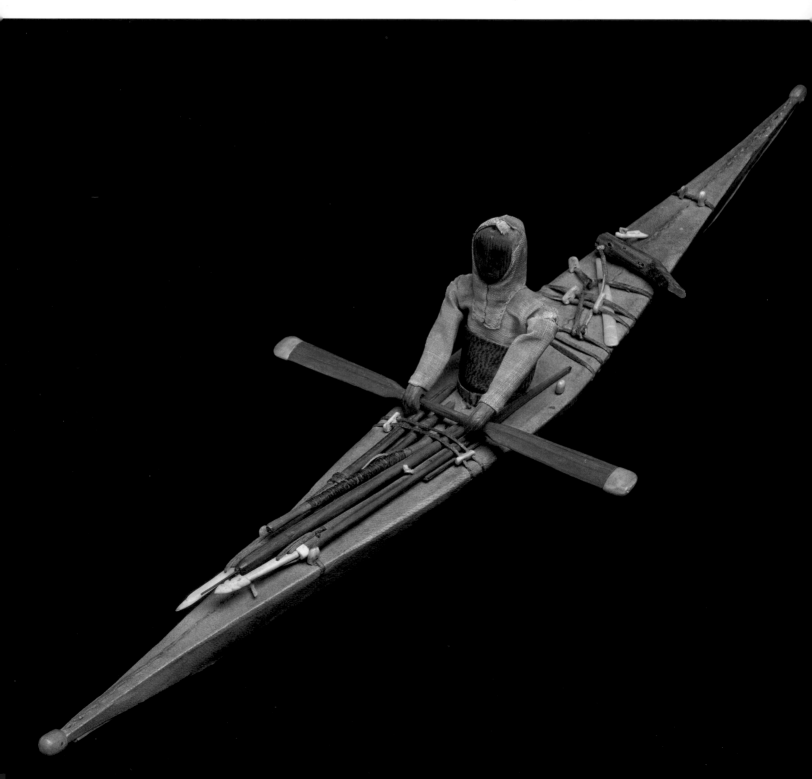

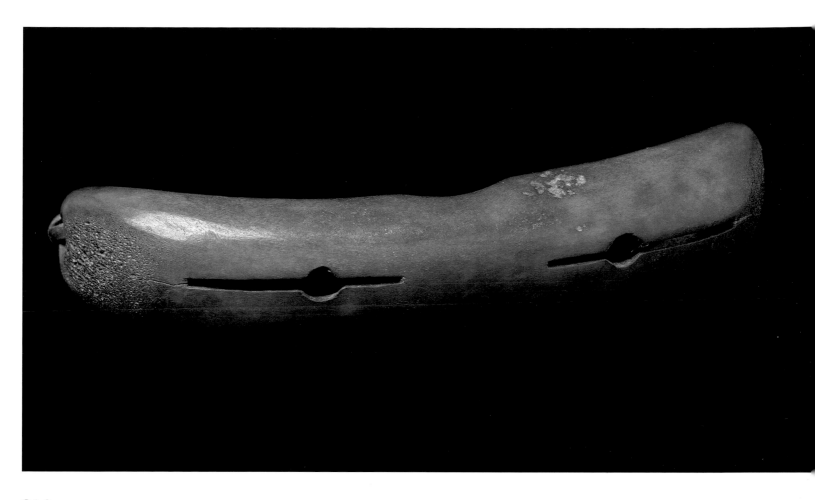

Cat. 2.
Point Hope, Alaska
**Snow goggles**
Collected 1905
Fossil bone, leather
Bequest of Frank C. and Clara G. Churchill;
46.17.9660

Cat. 21.
Simon Tookoome, Baker Lake, born 1934
*Nunamiutat*
1981
Linocut and stencil
Canadian Museum of Civilization Collection; BL 1981-031
Reproduced with the permission of Canadian Arctic
Producers

Cat. 22.
Mabel Nigiyok, Holman Island, born 1938
*I'll Go Plug the Weather*
1993
Stencil
Canadian Museum of Civilization Collection; HI 1993-020
Reproduced with the permission of Canadian Arctic
Producers

Cat. 27.
Bering Sea, Alaska
**Wooden kayak with man, paddle, fish,
and Palriayuk design**
Collected late 1930s
Wood, pigment, metal
Gift of the Estate of Corey Ford; 169.75.24874

Cat. 28.
**Harpoon with polar bear design**
19th or early 20th century
Wood, metal, ivory, skin
Found in the workroom of Vilhjalmur Stefansson
Stefansson Collection on Polar Exploration, Rauner Special
Collections Library, Dartmouth College Library
(detail)

Cat. 43.
King William Island, Terror Bay (Canada, Northwest Territories)
**Pair of caribou boots with liners**
Probably mid-20th century
Caribou fur, leather, cloth, woolen thread, (artificial?) sinew
Gift of Mrs. Lincoln Washburn, Class of 1935W; 174.2.25533

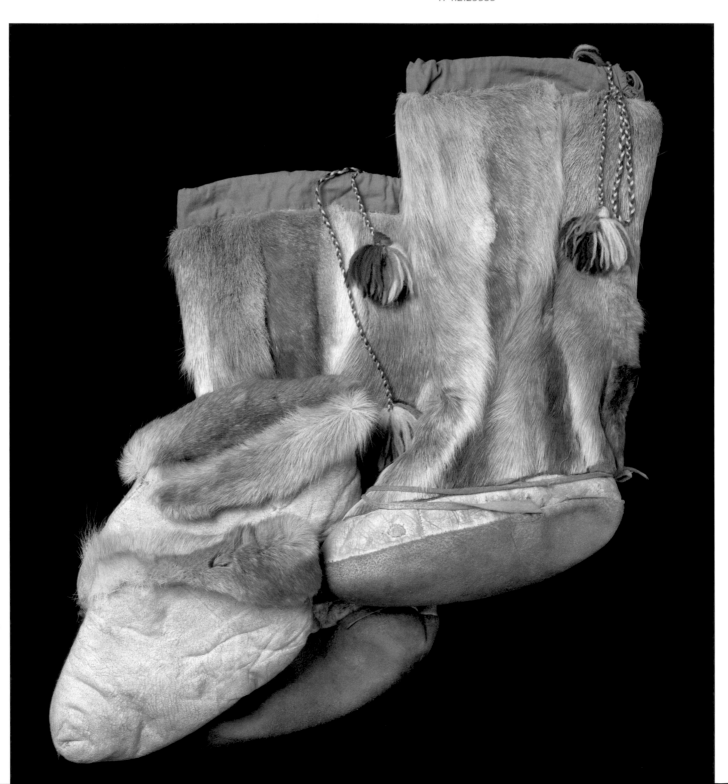

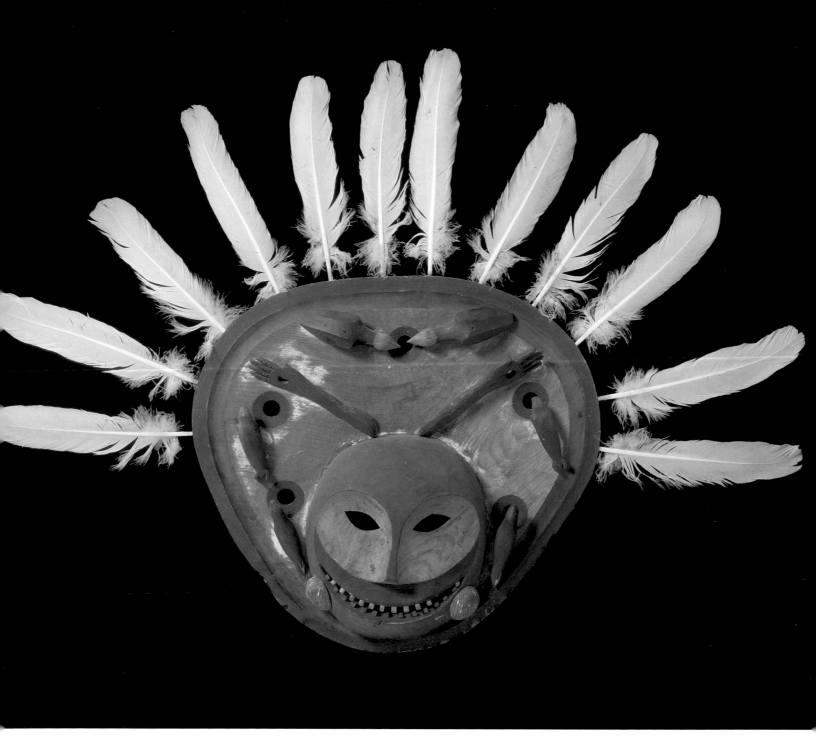

Cat. 29.
St. Michael, Alaska
**Shaman's *nepcetaq* mask**
Early 20th century
Spruce driftwood, shells, nails, rawhide suspension loop,
blue, red, and white paint, replacement feathers of seagull
(or turkey; originally swan or goose)
Gift of Lt. Col. Alfred T. Clifton, Class of 1927P;
42.15.7803

Cat. 65.
South Greenland
**Miniature of a hunting scene with blind and gun stand**
Collected 1938
Wood, paper, cloth, leather, seal fur and leather,
commercial thread, glue
Gift of George Murphy, Class of 1941; 39.70.7906

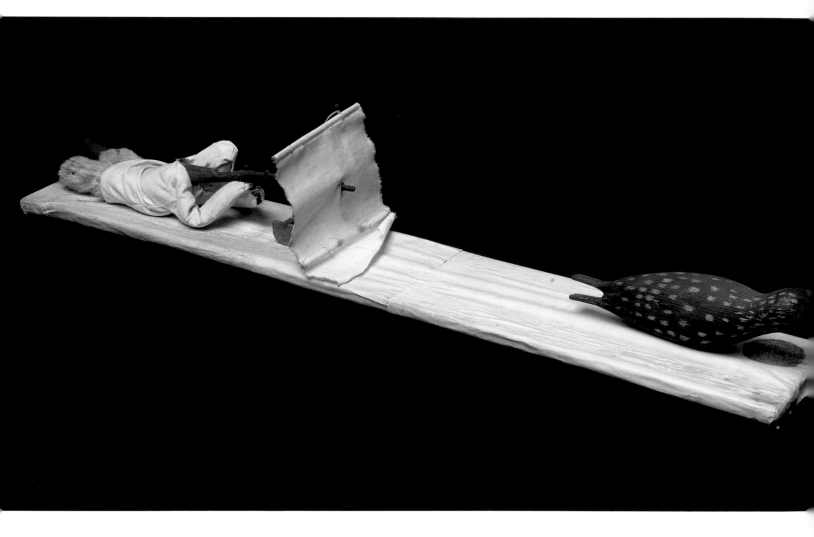

Cat. 44.
Tooksik Bay, Alaska
**Seal gut parka**
Mid-20th century
Seal intestines, cloth, thread, grass
Gift of Philip Nice and Cheryl R. Nice; 180.29.25951

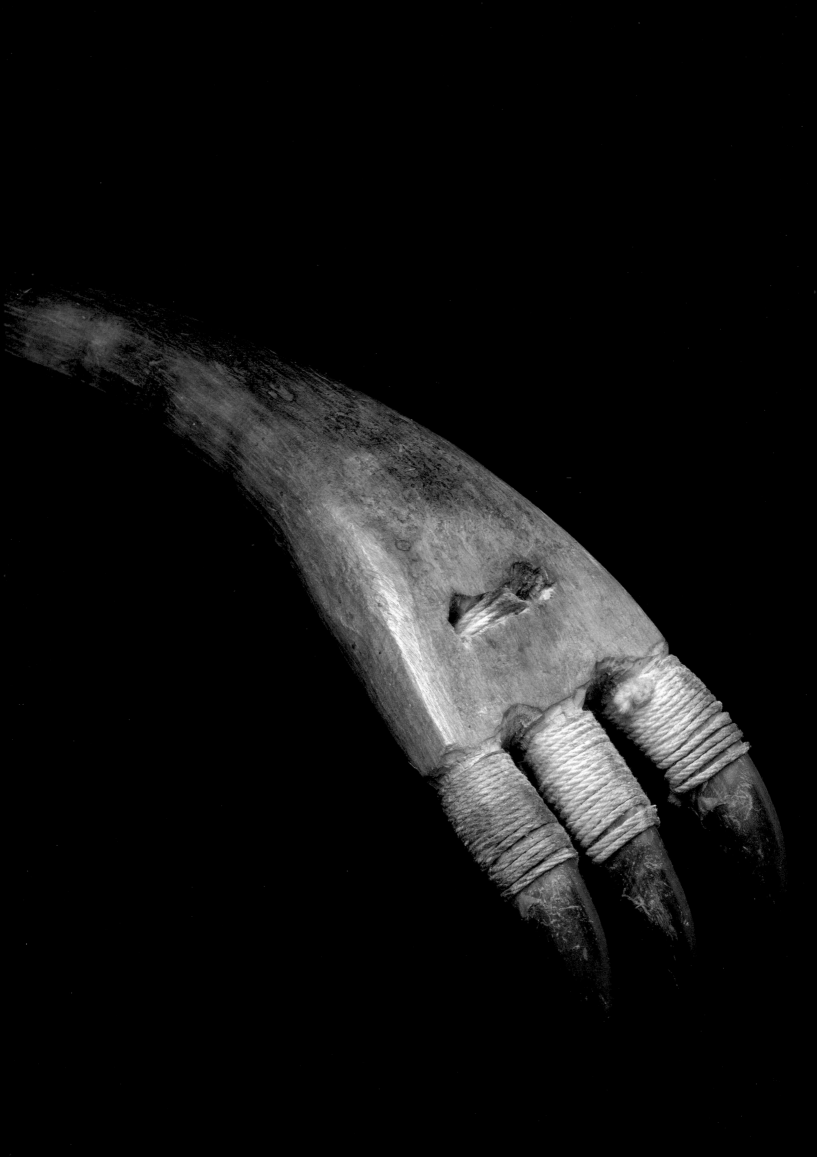

*Inuit spirituality is not just shamanism or Christianity. An important aspect of spirituality is to be able to provide for your family. Keeping them warm, clothed, and fed plays a vital part in one's identity as an Inuk. When a hunter is waiting for a seal, his mind cannot be in turmoil, as he has to be very patient. The seal is able to feel vibrations and in the winter can sense this if a hunter is at a breathing hole. The seal is also important to a woman's spirituality and to her contribution to the family. For example, there are different ways for a woman to prepare a sealskin.*

(Peter et al. 2002: 170)

## CATALOGUE

# Being Hunter—Being Game—Being Social: The Inuit in the Eastern Canadian Arctic

NICOLE STUCKENBERGER AND ERIK LAMBERT

he above quote, from a study by Ayu Peter and her fellow students at Nunavut Arctic College, shows that despite changes in the Inuit people's circumstances, their culture has changed less in the past century than outsiders might think. Most Inuit people continue to see themselves as part of a hunting society, and they formulate their self-perceptions in cosmological terms in relation to God, the land, weather, and animals: "The seal . . . provides us with more than just food and clothes. It provides us with our identity. It is through sharing and having a seal communion that we regain our strength, physically and mentally" (Peter et al. 2002: 167; see cat. 61).

In the past, as now, Inuit social life varied with the seasons. The Inuit people of the Canadian Arctic and Greenland lived together in groups of several families in a hunting camp on the sea ice in the winter, a time of great communal festivals.

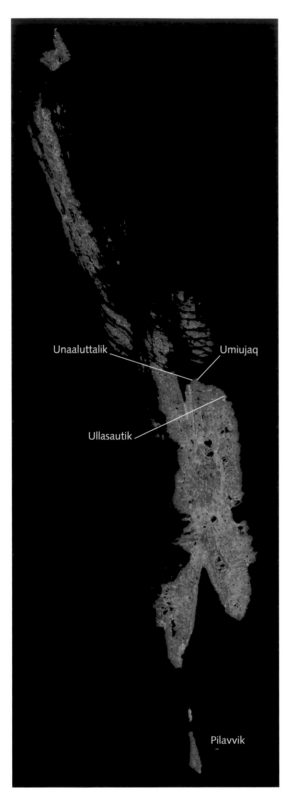

Unaaluttalik

Umiujaq

Ullasautik

Pilavvik

Fig. 16. The seasonal variations of the Arctic landscape (bottom of page) and a sample of Inuit place names of northern Belcher Islands, Hudson Bay, Canada; adapted from NASA/GSFC by Heather Carlos; place names and translation courtesy of Louis-Jacques Dorais.

### Samples of Place Names in the Belcher Islands

"Inuit place names relate to physical features of the land or to what people do on the land, and each place has its stories. They contain memories, information about how to use the land, and geographic orientation. And they link people to the land."

### Pilavvik (Where game is cut up)

"An island. It is called Pilavvik because it is the place where, a long time ago, they used to cut up the large walrus they killed there."

### Ullasautik (Traditional fox trap)

"The reason it is called Ullasautik is because a long time ago they built a huge trap for catching fox, made out of big rocks. It is called Ullasautik because it has an entrance made out of rock. It used to catch a lot of fox in the old time…These rocks are still visible today; one of them is a very big one…

"It was once used by a hunter who had lost his dogs, even if it belonged to another man. It is because it can work in that way that the place is called Ullasautik.

"The first ones who stopped using it left to go to another place where there was more game…They left it when they went to Kuujjuaq. This place was very much visited before there were Qallunaat (Europeans). It had no name in the beginning. It had no name for a long time, but when this *ullasautik* was built, it got its name from it."

### Unaaluttalik (Where there is Unaaluk [a person's name])

"A person died by shooting himself unintentionally, during fall, when the ground was already covered with snow, while they were living in tents and not in the snow houses yet. This place was inhabited by people, but only during the summer. When Unaaluk died, those who lived there were most probably the following families: Qautsaalik, Miiku, and Nappatuq."

### Umiujaq (Which looks like a boat)

"An island. It is said that it got its name because of the fact that it looks like a big boat…Since it is a high hill and an island, one cannot live there."

From *Inuksiutiit Allaniagait* 1, Quebec, Association Inuksiutiit, 1977, pp. 26–37.

## Hudson Bay, Canada

April 23, 2002

June 22, 2000

July 22, 2003

The camp dispersed into smaller family units during the summer. The identity of these various and changing groups was not established in relation to a larger social unit, such as a tribe, but in relation to the land they inhabited seasonally (fig. 16). A group's name is thus composed of a place name and the suffix -miut ("inhabitants of"). Qikiqtarjuarmiut, for example, would translate as "inhabitants of the large island." This type of name composition is still in use today.

Social relations, based on kinship and alliances, were often temporary, emerging from each camp member's work at a particular location. Great communal festivals were held in the fall or the winter, when the Inuit lived in the largest camps and the sun was about to disappear, so as to prepare for the return of animals by reestablishing proper relationships with all beings (see cat. 12). As one scholar points out, "For Inuit, success at the hunt was a result of respecting the soul of their quarry, of holding the proper attitude toward the seal. . . . The hunter is not extracting from the environment but creating a bond between his people and their environment" (Pelly 2001: 106). Although Inuit people today continue to live in settlements, they maintain a nomadic lifestyle and acknowledge the connection of animals to human social life and cosmology (see cats. 15–20).

Fig. 17. Tobias Ignatiussen, helped by kinfolk, builds a kayak. Photo by Cunera Buijs, RMC, Tasiilaq, 2001, courtesy of Rijksmuseum voor Volkenkunde, Leiden (The Netherlands).

The Inuit people "believed that the seal made itself available to the hunter, so that he could catch it. From the time that the seal gave itself, the hunter had an obligation . . . to share the seal with the people of his camp. If he failed to honour this obligation, the seal would not give itself to the hunter again . . . Sharing the seal ensured that there would always be seals to be caught" (Peter et al. 2002: 168). Social relationships were established and maintained by sharing game. Cooperation between men and women, and especially among hunters, was crucial (see cat. 48); alone, no one could survive. Hunters gained status by being generous providers of game. Women were also esteemed for skillfully transforming animal products into a meal to be shared, functional and beautiful garments for the family, and heat and light (figs. 17 and 18). Respect toward the seal was always part of these practices.

Inuit leaders were mostly men who possessed exceptional skill, knowledge, and wisdom, along with some years of experience, a strong personality, ownership of hunting gear, helpful family relations, and special access to spiritual powers. The isumataq was a "leader of camp life," while the angakut ("shaman") was the mediator between humans and non-humans. Although leaders could exert considerable influence, individuals were seen to be their own masters and responsible for their own decisions (Stevenson 1997: 233–38). However, those who behaved irresponsibly forfeited their credibility and were ignored—in the past, a very dangerous position to be in, as survival rested on cooperation (Oosten 1989: 337).

Living in the Arctic has always required specialized and well-adapted technology (see cats. 38 and 56). Before Western trade goods were available, the Inuit depended on animals for their food, shelter, clothing, and fuel (blubber was used for oil lamps and cooking). Animal bones, ivory, skins, and sinew provided utensils, tools, weapons, and spiritual objects such as amulets. From the earliest contacts with Europeans at various points in the eighteenth and nineteenth centuries, however, any new material or technology of Inuit or Qallunaat (European) invention was readily integrated into the culture if it made life in the Arctic easier or safer.[1]

Fig. 18. Two women sewing seal-leather cover for a kayak, Amassalik, May 1934. Photo by Van Zuylen, no. 4 AF 164, courtesy of Rijksmuseum voor Volkenkunde, Leiden (The Netherlands).

Depending on the region and season, a variety of food sources and materials for tools have always been used, such as bird's eggs, berries, seaweeds and other plants, and fish. Whaling was done from larger boats (*umiak*) along the ice edge in northwest Alaska. In the eastern Canadian Arctic, the smaller narwhal was pursued during its fall migrations. Sea mammals, pursued by boat in summer, at breathing holes in winter, and on the sea ice in spring, could supply most of the Inuit's nutritional needs. Land mammals such as caribou and polar bear, often hunted during the fall, provided the skins for winter clothing and additional food (Nuttall et al. 2005).

Hunting weapons were effective only when subjected to ritual observances that expressed the hunter's respect for his prey. They were made more powerful, however, by careful manufacturing, decorations that were attractive to the animals, and magical practices such as hunting songs (see cat. 28). Heavy harpoons with floats were used to capture and tire large sea mammals before they would be killed with a lance. Migratory birds were caught with lures, slings, and nets in spring. During the molting season in summer, kayakers used a bird harpoon to kill birds while they were less able to fly. Hunting migrating animals required an elaborate transportation and shelter technology, as well as social flexibility. In the winter a dogsled served hunters well; dogs were hardy, cautious with unsafe ice conditions, and able to orient themselves in bad weather (see cat. 6). Sled construction methods allowed for flexibility in the frame, which accommodated rough ground and ice conditions.

In the summer, men used kayaks for hunting. Kayak designs were adapted to the waves, the wind, and the behavior of animals, as well as carefully crafted to fit the body of the owner. Kayaks were fast, maneuverable, and light but also robust enough to allow the hunter to work even in rough seas (see cat. 62). The *umiak,* a larger boat that could be sailed or rowed, was used for moving camp or for whaling. Traveling in the summer could be especially difficult, as terrain could be muddy due to permafrost melt and rain.

Hunting was and continues to be essential to Inuit society. However, the awareness, intelligence, and agency of both animals and nature make the Arctic unpredictable and therefore dangerous for people on many levels. In response to human misconduct, animals might withdraw themselves or snowstorms might rage. In either case, an Inuit family that was unable to hunt could starve or freeze. Addressing especially the sea mammals at the center of Inuit hunting life, the many Inuit rules and seasonal rituals gave shape and some sense of control to their lives (see cat. 29).

The exhibition catalogue, like the exhibition, is divided into four sections that reflect the relation of the hunter to *sila,* to the animals that he seeks to harvest, to the ways in which hunting and prey contribute to the community, and to the technology used to hunt and to make goods from the game. A smaller section deals with different perspectives on climate change.

Unless otherwise noted, objects are from the collection of the Hood Museum of Art, Dartmouth College.

## NOTE

1. For a study on the introduction of GPS to the Inuit, see Aporta and Higgs 2005.

Cat. 30. St. Michael's area, Alaska, seal mask in the shape of a human face, collected about 1930s, spruce or cottonwood, paint (red, black), leather (once with fur). Gift of Lt. Col. Alfred T. Clifton, Class of 1927P; 42.15.7802.

**A hunter must know when to hunt and what equipment is appropriate to the season.**

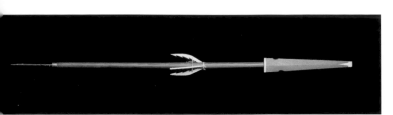

**1**

West Greenland, Disko Bay
**Bird spear with throwing board made by Karl Mathiasen in Sarkak** (see p. 5)
Mid-20th century
Wood, metal, ivory, thread
Gift of Per Jacobi; 160.12.14520

This spear's accompanying throwing board (*atlatl*) offers a hunter the leverage to throw it faster, farther, and more accurately. A sidearm flick of the wrist provides the necessary velocity to strike molting waterfowl from a kayak without capsizing into the frigid water. Three prongs with ensnaring barbs catch, rather than pierce, the prey, and a skilled hunter can trap multiple birds with a single launch. (Fitzhugh and Kaplan 1982: 74)

**2**

Point Hope, Alaska
**Snow goggles** (see p. 47)
Collected 1905
Fossil bone, leather
Bequest of Frank C. and Clara G. Churchill; 46.17.9660

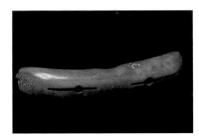

**3**

Spence Bay, base of Boothia Peninsula, northern Canada
**Snow goggles**
Possibly first half of the 20th century
Wood
Gift of Sherman O. and Anne L. Haight; 168.94.24455

Disorientation as a result of snow blindness can be deadly in the Arctic. Narrow horizontal slits in these driftwood snow goggles (*iggaak*) reduce glare and ultraviolet rays, especially in the spring, when ice and snow reflect the most sun. Bone, ivory, and sealskin are other common materials for goggles, though today polarized lenses have gained great popularity. (http://coas. missouri.edu/anthromuseum/pdfs/I_huntfish.pdf)

**4**

Aleut
**Visor/eyeshade**
Early 20th century
Wood, steel
Museum purchase; 54.67.13226

When pulled closely over the eyes in summertime, this visor keeps a hunter's vision unhindered. Its elongated front section is particularly effective for blocking out sun and sea spray while hunting in a kayak. (Varjola 1990: 176)

**5**

Pudlo Pudlat, 1916–1992, Canadian (Inuit)
***The Seasons***
1976
Lithograph (Pitseolak Niviaqsi, printer), Cape Dorset
Canadian Museum of Civilization Collection; CD 1976-061
Reproduced with permission of Dorset Fine Arts

Contrasting seasons dominate this print, which contrasts open water with ice, tents with igloos, and stars with the northern lights. Pudlo's interpretation emphasizes how human activities change as the weather does. In the lower left corner, the artist makes a personal confession. Angered by his sled dogs, who are caught on a rock, he snaps whips at them with both hands. "I am angry because my dogs won't listen to me," Pudlo admits. "I have just tangled them on the rock and I'm beating up my poor dogs . . . Dogs make the man, because a hunter can't go anywhere without them . . . With a snowmobile, you can't tell if it's going toward a dangerous area or not. With dogs, dogs can warn you." (Routledge and Jackson 1991: 131)

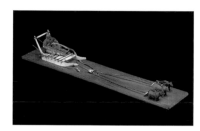

**6**
West Greenland
**Model dog sled, team of three dogs, and driver**
Collected about 1897
Wood, fur, leather, bone, thread
Gift of Mrs. William Stickney, wife of William Stickney, Class of 1900;
164.21.15445

This familiar Arctic hunting scene reflects the importance of dogs in the far north. Their companionship and keen senses provide more to a hunter than just reliable transportation. A smart dog can smell seawater from a distance, revealing dangerous cracks in the ice or helping to find seals' inconspicuous breathing holes (*agluit*). (Pelly 2001: 46–49)

**7**
Southwest Greenland
**Miniature one-person kayak with equipment and figure**
Mid-20th century
Sealskin, wood, ivory, beads, iron
Gift of David Nutt; 159.9.14390

The float board (*asaluuk*) sitting above the deck of a Greenlandic kayak holds the line for a hunter's harpoon. Kept both tidy and safe from being washed away, the harpoon line sits coiled atop the board, allowing space for other tools below, including the harpoon itself. In action, the line will pay out quickly to follow a harpooned seal. (Heath and Arima 2004: 18, 23; Zimmerly 2000: 57)

**8**
Greenland
**Miniature kayak with paddler and equipment** (see p. 46)
About 1920
Hide, wood, fabric, bone
Purchased through the Hood Museum of Art Acquisitions Fund; 2005.56

Wind at any speed can be a nuisance in lightweight seal- or caribou-skin kayaks. In these conditions, paddlers often slide their windward grip down to the blades of the flat, symmetrical paddles (*paatit*) to decrease wind resistance. (Heath and Arima 2004: 46, 151)

**9**
West Greenland
**Miniature one-person kayak**
Collected about 1897
Sealskin, wood, ivory, metal pieces
Gift of Mrs. William Stickney, wife of William Stickney, Class of 1900;
164.21.15434

Adapted to the frequent gales along Greenland's Arctic coast, the local kayaks are long (seventeen to eighteen feet), narrow (nineteen inches), and low in profile. This sleekness also proves useful for hunting, as seals are less prone to spot the watercraft. (Heath and Arima 2004)

This section follows the travel, killing, and ritual observances involved with hunting and examines the relationship between hunter and prey.

**10**

Alaska

**Harpoon decorated with seal carving; point missing**

Collected about 1897

Ivory, wood

Gift of Timothy Sample; 54.71.13241

Understanding a seal's soul is just as important to a hunter as his tools and skills. Knowing that seals are intelligently aware, this hunter has engraved a beautiful seal's body into his harpoon's head to show respect. The carving will attract the seal and permit the hunter an easier kill.

**11**

South Greenland

**Hunting blind to camouflage a hunter**

Collected about 1938

Wood, cotton cloth, twine, leather lines

Gift of George Murphy, Class of 1941; 39.70.7907

Hunting in the Arctic requires ingenuity, because the icescape provides little camouflage and seals are alert. One common method of hiding from one's prey is the hunting blind, which slides along the snow or ice on runners and conceals the form of the hunter—and his gun—from the seal. Polar bears use the same technique, covering their black noses with their paws when stalking a seal.

**12**

Alaska

**Ice scratcher (see p. 56)**

Collected late 1930s

Wood, seal claws, twine

Gift of the Estate of Corey Ford; 169.75.24830

Seals resting on the ice periodically wake to check for predators. Anticipating this routine, a hunter will crouch behind one of his large white mittens, crafted from white polar bear or dog skin, and scrape the ice with his scratcher (*cetugmiarun*). The rough sound pacifies the prey by mimicking the noise of another seal creating a breathing hole and thus allows the hunter to advance. (Fitzhugh and Kaplan 1982: 78)

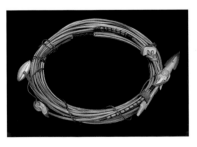

**13**

Alaska

**Harpoon line with attachments**

Collected late 1930s

Bearded seal skin, bone, thread

Gift of the Estate of Corey Ford; 169.75.24923

When a seal is speared, the hunter's harpoon line connects him to the animal both literally and spiritually. This line, made from bearded seal hide, has an ivory toggle in the form of a seal and an ivory line attachment that joins the harpoon to the coil of thong. Sometimes harpoons or line express reverence through symbols or animal shapes, but the markings on the tip and the toggles here probably identify the owner instead. (Fitzhugh and Kaplan 1982: 76; Elmer Harp, interviewed by Erik Lambert, August 6, 2006)

**14**
Unalakleet, Alaska
**Tom cod fishing equipment**
Collected 1948–57
Wood, ivory, red thread, green string, metal
Gift of Sally Carrighar; 166.43.16171

In the morning during winter or spring, men chop through as much as six feet of ice to make a fishing hole. Women will then release the sinker, lure, and hook through the opening and into the frigid water. By knocking a second, pointy stick against the rod, the lure jigs through the water, encouraging tom cod to take the bait.

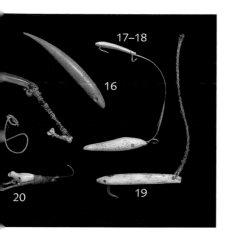

**15**
Point Barrow, Alaska
**Fishing hook**
Collected 1905
Iron, ivory, baleen, bead
Bequest of Frank C. and Clara G. Churchill; 46.17.9648

**16**
Nome, Alaska
**Fish-shaped sinker**
Collected 1948–57
Ivory, metal
Gift of Sally Carrighar; 166.43.16161

**17–18**
Alaska
**Fish hook-lure, sinker**
Collected late 1930s
Ivory, metal
Gift of the Estate of Corey Ford; 169.75.24917

**19**
Point Barrow, Alaska
**Fish hook-lure**
Collected about 1905
Ivory, sinew, iron, copper (?)
Bequest of Frank C. and Clara G. Churchill; 46.17.9646

**20**
Canada
**Fish hook with carved seal (see cover)**
Late 19th–early 20th centuries
Ivory, brass or copper, sinew, fishing line
29.58.7934

Fine equipment prepares men and women for an outing on the sea ice mentally, physically, and spiritually. Their prey also appreciates this craftsmanship, which tells the animal spirits that these people have prepared for their ocean hunt. By venerating seals, the greatest fishers of all, with carved likenesses, hunters enhance the charm of their hooks and sinkers.

**21**
Simon Tookoome, Baker Lake, born 1934
*Nunamiutat* (see p. 48)
1981
Linocut and stencil
Canadian Museum of Civilization Collection; BL 1981-031
Reproduced with the permission of Canadian Arctic Producers

Humans and animals, woven together into the fabric of the universe, harmoniously intermingle as *nunamiutat,* or "those who dwell on the land." Integrated into this summer landscape, which centers on a melting lake, they coexist as part of land, water, and sky.

**22**

Mabel Nigiyok, Holman Island, born 1938
*I'll Go Plug the Weather* (see p. 49)
1993
Stencil
Canadian Museum of Civilization Collection; HI 1993-020
Reproduced with the permission of Canadian Arctic Producers

Calling upon the strong spirit of the polar bear, a female shaman beats a drum to pacify the weather spirit, Narssuk, a giant baby who emits bodily wind when humans misbehave. The result, seen here, is a violent windstorm that troubles hunters and their families. Created for an art market with a great demand for shamanic and mythological references, this stencil highlights traditional concepts in a contemporary context.

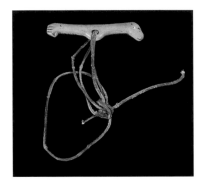

**23**

Point Barrow, Alaska
**Seal drag with carved and engraved handle**
Collected 1905
Ivory, rawhide
Bequest of Frank C. and Clara G. Churchill; 46.17.9651

After spearing a seal, the hunter transports the heavy load (up to five hundred pounds) with a seal drag (*sapaniaq*). A line of walrus rawhide loops through an incision cut in the seal's lower jaw and connects to an ivory pulling handle. Using the frictionless ice, the hunter can drag the seal to his home or dogsled. The handle's decoration honors the slain animal and reminds seal spirits of the necessity of the hunt. (http://www. dfo-mpo.gc.ca/zone/underwater_sous-marin/hseal/seal-phoque_e.htm)

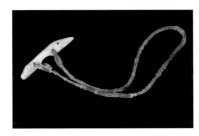

**24**

Point Barrow, Alaska
**Seal drag with carved and engraved handle**
Collected 1905
Ivory, walrus rawhide
Bequest of Frank C. and Clara G. Churchill; 46.17.9650

Alaskan hunters revere seabirds as excellent predators that hold great power in two worlds, sea and air, despite their small size. This drag handle, one side of which is carved into a seabird, reflects the hunter's belief in the connection between seals and seabirds. After death, a seal's soul concentrates in its bladder, and a hunter can spiritually connect with the animal through the link of the drag handle. (Fienup-Riordan 1990: 27)

**25**

Greenland
**Miniature one-man kayak, fully equipped** (see p.11)
Collected 1950
Sealskin, gut, ivory, wood, steel
Gift of Peter S. Dow; 50.14.12412

The tidy setup of a kayak is not only essential for successful hunting but also reflects a hunter's preparation and skill and expresses respect for prey, who appreciate an experienced hunter.

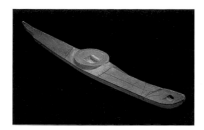

**26**
Nunivak Island(?), Bering Sea, Alaska
**Miniature kayak with Palriayuk design**
Collected late 1930s
Wood, paint
Gift of the Estate of Corey Ford; 169.75.24873

The likeness of Palriayuk, a powerful water monster and extraordinary hunter, embellishes the curving shape of this kayak miniature. Likenesses or paintings of Palriayuk on the side of a kayak bring the monster's powers to the paddler, imbuing him with great fortitude and skill. (Curtis 1927, in Zimmerly 2000: 40; Edwin T. Adney and Howard I. Chapelle, *The Bark Canoes and Skin Boats of North America* [Washington, D.C.: Smithsonian Institution, 1964], 199ff)

**27**
Bering Sea, Alaska
**Wooden kayak with man, paddle, fish, and Palriayuk design** (see p. 50)
Collected late 1930s
Wood, pigment, metal
Gift of the Estate of Corey Ford; 169.75.24874

The hands of an *angakok* (shaman) can gather animals from across the worlds. In Yup'ik culture the oversized hand, which embodies the spirits' ability to release animals to the human world, connects the hunter to the spiritual world and its riches. This kayaker caught a large fish and displays a magical hand reaching out to the sea. (Fitzhugh and Kaplan 1982: 109; Fienup-Riordan 1996: 166, 196)

**28**
**Harpoon with polar bear design** (see p. 51)
19th or early 20th century
Wood, metal, ivory, skin
Found in the workroom of Vilhjalmur Stefansson
Stefansson Collection on Polar Exploration, Rauner Special Collections Library, Dartmouth College Library

Bearing his teeth on the harpoon's shaft is a polar bear charm (see page 51); less obvious is another bear likeness carved into the bone foreshaft. One of the most formidable dangers on the sea ice and on land is the polar bear, who demands equal measures reverence and fear. The carving uses the power of the bear, while the pick at the base is used to test the ice's thickness.

**29**
St. Michael, Alaska
**Shaman's *nepcetaq* mask** (see p. 53)
Early 20th century
Spruce driftwood, shells, nails, rawhide suspension loop, blue, red, and white paint, replacement feathers of seagull (or turkey; originally swan or goose)
Gift of Lt. Col. Alfred T. Clifton, Class of 1927P; 42.15.7803

Each Yup'ik shaman designs and oversees the construction of his mask (*nepcetaq*). The seal breathing holes surrounding the face represent passages between the spirit and human worlds. Sea mammals travel through these holes to the human realm to be hunted, and a powerful shaman can travel through them to other realms. The feathers also link the realms. The shaman summons the *nepcetaq*'s power to increase hunting success or to tell the future, especially regarding life and death. (Fienup-Riordan 1996: 77–78)

**30**

St. Michael's area, Alaska

**Seal mask in the shape of a human face** (see p. 61)

Collected about 1930s

Spruce or cottonwood, paint (red, black), leather (once with fur)

Gift of Lt. Col. Alfred T. Clifton, Class of 1927P; 42.15.7802

Male-associated "smiles" and female-associated "frowns" frequently adorn Yup'ik masks. "Masking goggles" rings draw attention to the eyes and potentially mark the purpose of the mask—to see into another realm. The eye rings may also represent the spirits of seals or otters. (Zimmerly 2000: 40; Fienup-Riordan 1996: 166)

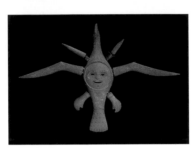

**31**

Nunivak Island, Alaska

**Loon mask**

Collected late 1930s

Spruce wood, feathers, black and red paint

Gift of the Estate of Corey Ford; 169.75.24909

In Yup'ik mythology, loons have restorative powers often connected with healing blindness. They also communicate with people, as Yup'ik elder Alma Keyes recounts: "They say when someone is going to die, the loons stay around and continue to dive in and out of the water . . . Loons are quite aware. And they know when things move around. They would tell us to be alert when we pick berries. When other animals are nearby the loons don't keep quiet. They scream out and sing" (Fienup-Riordan 1996: 241). Along with owls and other sea birds, loons also often embody the helping spirit (*tuunraq*) of a shaman, who probably lent this mask its specific origin and meaning. (Fienup-Riordan 1996: 59–60, 217, 240–41, 246; http://www.tribalarts.com/feature/riordan/)

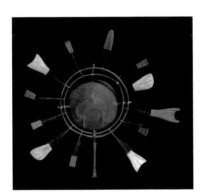

**32**

Nunivak Island

**Walrus mask (see title page)**

Collected late 1930s

Wood, gold, red, black paint, sinew, black feathers

Gift of the Estate of Corey Ford; 169.75.24910

Although every mask is unique, some themes and designs repeat because of their cosmological origins. For example, the arrangement surrounding this walrus's features—concentric circles adorned with bird feathers and seal flippers—are common to many Yup'ik masks. Yup'ik elder Dick Andrew relays a dance performance with walrus masks at Kayalivik, Alaska: "Ones that had walrus masks would make sounds like walrus [during the dance] . . . They would talk about the fish mask and say they were presenting it hoping to receive [more fish] . . . They would make masks that represent their *tuunrat* . . . And in the spring there would be lots of walrus if one of them had presented a walrus mask." (Fienup-Riordan 1996: 103)

**33**

Josephee Kakee, Inuk, 1911–1977

***Tornrak Chief***

1964

Pangnirtung

Canadian Museum of Civilization Collection; IV-C-5847

Gift from the Collection of James and Alice Houston and the American Friends of Canada, 1999

Reproduced with the permission of Canadian Arctic Producers

*Tornrak Chief* recalls the story of Angmarlik, a famous whaling captain, shaman, and camp leader who sought advice about weather conditions and whale movements from two other shamans, one of night and one of day. This shaman transforms into an animal, perhaps a polar bear, to accommodate Angmarlik.

**34**

Simon Tookoome, Baker Lake, born 1934
*The World of Man and the World of Animals Come Together in the Shaman*
1974
Stonecut and stencil print
Canadian Museum of Civilization Collection; BL 1974-012
Reproduced with the permission of Canadian Arctic Producers

This shamanistic transformation from a person into a bird relates man and animal to each other; humans and other creatures share the same substance, and shamans retained the mythological ability to pass between the two states of being. The hoof of a caribou and shapes of dogs also animate this figure, whose face is formed of two different profiles.

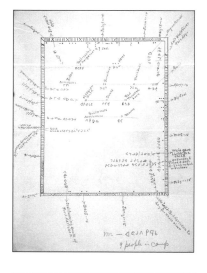

**35**

**Inuk calendar diary**
May 18, 1914–April 18, 1915, possibly kept by
Ang-oo-tik-ek at Port Harrison and Whale River
Hudson's Bay Company Archives, Archives of Manitoba; Ref. E.128/1
Reproduced with the permission of Hudson's Bay Company Archives

This calendar diary's borders show the days and months, with notations of important events and catches written in Inuktitut and then translated by George Robert Redfearn, a trader. A calendar diary also displays the movement of people in conjunction with seasonal patterns. As Christian belief spread, traditional rules governing hunting were integrated into the new framework, which included resting on the Sabbath. The crosses thus marking every seventh day reveal how Christian practices entered this tradition-bound cosmology. (MacDonald 1998: 204–5; Hudson Bay Archives object sheet)

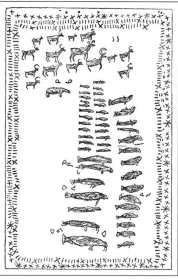

**36**

**Reproduction of the calendar of Kroonook**
No original exists (Revillon Freres 1923: 21)

Visually representing the days on which the hunter captured game, this calendar diary depicts the sizes and shapes of each animal he caught. Artists today likewise try to provide an accurate picture of their own life experiences.

# SECTION 3. BEING SOCIAL

This section presents what happens when the dead animal is brought back to the community. In the house of the hunter, it is the homemaker's task to take care of the household and community by processing and distributing the meat and sewing garments. The body of the seal provides the medium through which social relationships are established and maintained.

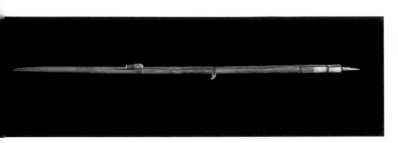

**37**
Alaska
**Walrus harpoon with a small ornamentation**
Early 20th century
Wood, ivory, hide
Gift of Timothy Sample; 54.71.13242

Slaying a one-ton beast is no easy task, but a long-range harpoon like this one will injure and then tire the walrus as it tries to drag the shaft sideways through the water. Eventually the hunter can approach close enough to kill it with a short-range spear or club. The harpoon foreshaft's length is based on the walrus's typical blubber thickness. Beyond meat and hunting materials, the walrus also provides the community with oil for heat and cooking. (Rousselot 1994: 54; Fitzhugh and Kaplan 1982: 80–81; http://alaska. fws.gov/fisheries/mmm/walrus/nhistory.htm)

**38**
Bathurst Island
*Ulu*
Collected mid-20th century
Bone, horn, copper, iron
Gift of Sherman O. and Anne L. Haight; 168.94.24462

Women use the semilunar-shaped all-purpose knife (*ulu*) to process and prepare seals and fish. Smaller versions prove handy for sewing and cutting out patterns for clothes. The design of each *ulu* is unique and depends on how the user prefers to hold the tool. Men use long knives instead. (Fienup-Riordan 2005: 173; Issenman 1997: 14)

**39**
Baffinland
**Lamp (*qulliq*)**
Collected about 1929–30
Soapstone
Gift of Robert O. Fernald, Class of 1936; 173.32.25508

The lamp (*qulliq*) was the glowing center of a household. Wicks, often made from moss, burned seal or walrus oil to provide light and heat; the blackening here suggests cooking applications as well. The small hole drilled in the center indicates that the owner may have "killed" the lamp in a ritual to disconnect her spirit from the object. The *qulliq* has largely given way to electric lights, central heating, modern kitchen stoves, and camping burners. Sometimes, at special occasions such as a graduation, an older woman might light a *qulliq* to celebrate. (Fienup-Riordan 2005: 187)

**40**
Point Barrow, Alaska
**Ladle/horn spoon**
Collected 1905
Mountain sheep horn
Bequest of Frank C. and Clara G. Churchill; 46.17.9662

This perforated ladle, formed from the horn of a Dall sheep, is a cooking and eating utensil made to take advantage of the shape of the horn. (http://www. pc.gc.ca/canada/nature/archives/2003/av-ap/archives1_E.asp; Rousselot 1994: 46)

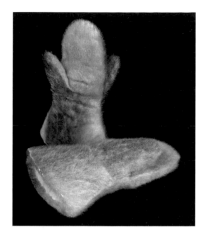

**41**
West Greenland
**Pair of sealskin mittens, double-thumbed** (see p. 20)
1897
Sealskin and fur lining (probably dog fur)
Gift of Mrs. William Stickney, wife of William Stickney, Class of 1900;
164.21.15432

An excellent example of form following function, the extra thumbs on these watertight mittens (*maattaalit*) keep paddlers' hands dry. In the summer, flipping the mittens dries the waterlogged side; when temperatures are colder, the hunter can scrape away built-up ice while maintaining his grip. The outside fur is unusual for mittens of this type. (Buijs 2004: 53)

**42**
Point Hope, Alaska
**Pair of waterproof boots** (see p. 6)
Mid-20th century
Seal and probably bearded seal leather, sinew, cord, thread
Gift of Alan Cook, Class of 1960; 163.47.15161

These sealskin boots (the fur has been scraped away) are best in late spring and early summer. Despite the advent of rubber galoshes, they remain common in the Alaskan and Canadian Arctic, as does their detailed sewing process. To avoid penetrating two layers of hide in any one place, boot seamstresses begin with a running stitch that dips into the inside skin without piercing it completely. Pinching an extra layer of the sole, the seamstress then pulls together the now-tight seam with a second threading to create the double-fold waterproof stitch (*ilujjiniq*). (Issenman 1997: 90–91)

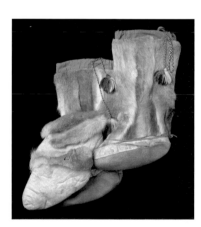

**43**
King William Island, Terror Bay (Canada, Northwest Territories)
**Pair of caribou boots with liners** (see p. 52)
Probably mid-20th century
Caribou fur, leather, cloth, woolen thread, (artificial?) sinew
Gift of Mrs. Lincoln Washburn, Class of 1935W; 174.2.25533

With their caribou fur liners, these boots (*kamiit*) warm the feet of winter travelers. A skillful seamstress sews the caribou fur, cloth, and other skins with wool thread and sinew, or modern materials such as dental floss. Even the smallest details are important; for example, the tiny hairs on the outer soles provide valuable traction on slippery snow and ice.

**44**
Tooksik Bay, Alaska
**Seal gut parka** (see p. 55)
Mid-20th century
Seal intestines, cloth, thread, grass
Gift of Philip Nice and Cheryl R. Nice; 180.29.25951

Sealgut parkas are watertight to keep the fur layers worn underneath dry during kayaking. To create such thin garments, which weigh between twenty and twenty-five grams, women scrape and wash the insides of seal intestines, blow them up and tie them off, then hang them to dry. Two days later they cut the balloons lengthwise into strips (*apuilitaq*), which they can then sew into a parka. (Rousselot 1994: 34; Issenman 1997: 81–82; Varjola 1990: 267)

**45**
**Caribou parka with caribou pants, canvas over-pants, and caribou mittens** (see p. 16)
1913–18
From Canadian Arctic Expedition: possessions of James Crawford
Stefansson Collection on Polar Exploration, Rauner Special Collections
Library, Dartmouth College Library; Stef-Realia 213

This caribou parka, second in warmth only to polar bear fur, has intricate hand-sewn stitching. The canvas over-pants, crafted partly on a sewing machine, exhibit very different seams. All garments would have been custom-fitted.

**46**
Southwestern Greenland
**Miniature *umiak* with four women and one man, a dog, and camping equipment** (see p. 28)
Collected prior to 1955
Sealskin, wood, cloth, metal pieces, hair(?)
Gift of David Nutt; 159.9.14387

Much larger than a kayak and capable of carrying up to twenty people, an *umiak* is an open boat often powered by women with single-bladed paddles. The cache of women's gear—*uluit* and cooking materials—reinforces the *umiak*'s female association. To this day, its capacity makes this boat, like the motor canoe, very useful for scouting, or moving camp in the summertime. (http://scaa.usask.ca/gallery/northern/content?pg=ex13-3; Rousselot 1994: 50)

**47**
Alaska
**Miniature *umiak* with sail and oars**
Collected late 1930s
Gift of the Estate of Corey Ford; 169.75.24931

In Alaska men used to paddle or sail *umiaks* while hunting for whales. The boat's size and hull construction offer greater protection than a kayak when approaching large game. Stretched and sewn over the stout beams are the tough, elastic skins of harp or bearded seals, or walrus hides. (Rousselot 1994: 50)

**48**
Kodiak Island, Alaska
**Kayak (*baidarka*) miniature with three men, two harpooners and one on a paddle** (see p. 27)
Collected mid–19th century
Skin, wood, paint, thread
13.1.591

The traditional three-man kayak (*baidarka*) answered the requirements of the fur trade in the Bering Sea, frequently serving as a taxi service for missionaries and Russian traders. Storing furs and other goods in the kayak's cavity, these passengers traveled long distances to spread ideas or barter valuable goods. The *baidarka* also carried small whaling teams, as here.

This section concludes the cycle of hunting with the production, improvement, and innovation of hunting and household tools and the teaching of knowledge and skills that prepare the hunter, the spouse, and the next generation for a new day out on the land and in the settlement.

**49**

North Greenland
**Harpoon and spear thrower**
Collected about 1938
Wood, ivory, hide, iron
Gift of George Murphy, Class of 1941; 39.70.7903

The distance a harpoon can be thrown depends on the location of its balancing point and on the hunter's skill with the throwing board (*atlatl*). In a successful strike the ivory toggling head at the point of the harpoon will enter the flesh and twist ninety degrees, locking fast into the prey. The seal or walrus will dive immediately, pulling the harpoon away from the hunter. Large and tenacious sea mammals sometimes require a second strike with a harpoon or spear. (Rousselot 1994: 54; http://pwnhc.learnnet.nt.ca/exhibits/nv/harpoon.htm)

**50**

Mary Ashevak Ezekiel, Cape Dorset, born 20th century
***Woman Sewing Skin Boots***
1960
Print
Purchased through the Guernsey Center Moore 1904 Memorial Fund; PR.961.113

Having cut foot-sized shapes from a slab of seal fur, this woman is sewing together a new pair of sealskin boots. Holes around the sides of the unused fur indicate that she stretched and dried the skin earlier by tying its edges to a rack. She needs only a few tools for the entire process, which she keeps nearby.

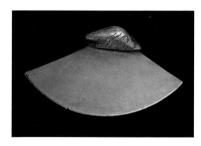

**51**

Alaska
***Ulu***
19th century, collected late 1930s
Ivory or bone handle, metal blade
Gift of the Estate of Corey Ford; 169.75.24886

The blades and handles of the all-purpose knives (*uluit*) that women have used since before recorded history have gradually changed over time. Once made from slate or stone, blades are now usually steel sawblades or other scrap metal. New materials or technologies that make things work better are welcomed in Inuit culture. (Fienup-Riordan 2005: 173; Issenman 1997: 16)

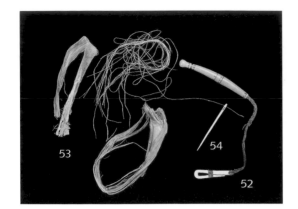

### 52

Point Barrow, Alaska
**Needle case with a belt hook**
Collected 1905
Ivory, leather, cloth
Bequest of Frank C. and Clara G. Churchill; 46.17.9676

Carving a hole into ivory or extracting marrow from bone produces a hollow cylinder for storing sewing needles. In general, older needles are also made from ivory or bone, whereas newer ones are steel, a material that is embraced today. A leather strip holds the needles safely within the cylinder, and extracting a needle only requires a slight tug on the ball-shaped end. The hook at the opposite end attaches to a woman's belt. (Issenman 1997: 33, 231)

### 53

Alaska
**Two strips of Caribou sinew thread**
Collected about 1960
Gift of Alan Cooke, Class of 1960; 163.47.15155

When a hunter brings a caribou back to camp, a seamstress slices its dorsal sinews (*uliut*) from the back, cuts them into ribbons, then cleans them and hangs them to dry. After a day, she splits the ribbons into thin strands with her teeth or thumbnail. She can store this sinew thread (*ivalu*) or use it immediately to sew clothing, tents, bedding, or kayaks. Today, seamstresses often use artificial sinew or waxed dental floss as well. (Issenman 1997: 84)

### 54

North Greenland
**Needle**
Collected 1938–39
Bone
Gift of George Murphy, Class of 1941; 39.70.7880

Older women accustom girls to sewing, a vital part of countless crafts, at an early age. A Yup'ik elder named Cungauyar Annie Blue recalled, "A young girl's lessons in sewing begin with sewing unraveled seams or old, torn parkas. Some girls discovered that they enjoyed sewing early in life, and some did not." (Fienup-Riordan 2005: 239)

### 55

Point Hope, Alaska
**Skin scraper**
Probably 19th century
Stone (nephrite blade), carved wooden handle
Gift of Glover Street Hastings III; 181.2.26095

Smoothed and grooved specifically to fit its user's hand comfortably, each skin scraper (*sakuuti*) varies greatly in contour. Various scrapers are designed to stretch and soften skins, de-hair hides, or remove flesh, fat, and oil from hides (as this scraper does). (Issenman 1997: 64)

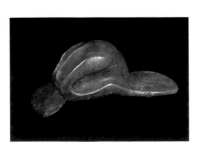

### 56

Couper Island, northeast of mouth of Coppermine River, Coronation Gulf
**Scraper**
Collected 1965
Wooden handle, recycled metal sheet, screw, rope
Gift of Elmer Harp Jr.; 165.38.15682

Finding the parts and then having the vision to integrate them into a functional tool is a challenging task. An ingenious design wrought from recycled, low-cost materials, this skin scraper (*sakuuti*) embodies the virtue of resourcefulness.

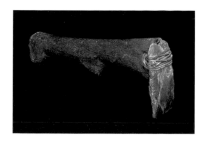

**57**
Alaska
**Stone adze blade hafted to carved wooden handle**
19th century
Wood, jade, rawhide
Gift of Glover Street Hastings III; 181.2.26093

A blade like this, crafted from jade found along the talus slopes or valleys of the Kobuk River in northwestern Alaska, would have been used to split and hollow wood. At one time, it may have been used for traditional kayak building. An adze blade (*kepun*) is fastened tightly to a contoured handle by sinew or, in this case, rawhide. This particular object may have been made for the tourist trade. (Varjola 1990: 249; Fienup-Riordan 2005: 169; Rousselot 1994: 63; Sturtevant/Damas 1984: 290; http://www2.nature.nps.gov/geology/parks/kova/index.cfm)

**58**
Nome, Alaska
**Drill (shaft and point)**
Early 20th century
Wood, steel, brass
Gift of Lt. Col. Joseph W. A. Whitehorne III; 160.56.14633

**59**
Alaska
**Mouthpiece for a drill**
Collected late 1930s
Wood, two nails, stone piece to accept the drill
Gift of the Estate of Corey Ford; 169.75.24889

59

**60**
Alaska
**Drill bow**
Collected late 1930s
Ivory and leather
Gift of the Estate of Corey Ford; 169.75.24889

A drill kit (*igauquk*) was used to inscribe ivory, start a fire, or create holes. By clamping the drill cap in one's teeth and vertically inserting the shaft between the surface and the cap's friction-reducing stone, the bow can swivel the shaft and bit, which digs into the surface. This bit is probably a carpenter's nail, but before steel, bits were made from stone and required drilling from opposing sides. To perform similar tasks today, craftsmen usually prefer power drills. (Fitzhugh and Kaplan 1982: 169; Rousselot 1994: 62)

**61**
Alaska
**Two figures carrying a seal**
Early 20th century
Soapstone
Purchased through the Julia L. Whittier Fund; S.957.27

Straining from the heavy load, these two hunters work together to carry a harp seal, a hard but happy task at the end of a successful hunt.

**62**
Alaska, Inuit
**Miniature Bering Strait–type kayak with rack for harpoon line**
Collected late 1930s
Skin, wood
Gift of the Estate of Corey Ford; 169.75.24887

The design and construction of one's own kayak is based on the lengths of one's arms and even fingers. Just as the boat fits the person perfectly, its components in turn fit the boat. The wooden structure on the foredeck is a float-board or drag (*asaluuk*), which also serves as a rack for harpoon line. (Zimmerly 2000: 37, 57; Buijs 2004: 145)

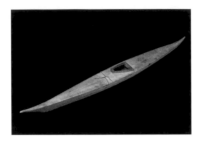

**63**
Greenland
**Miniature kayak with unusual triangular cockpit**
About 1924
Wood, sealskin, nails, iron wire
Gift of Paul and Lei Goddard; 182.3.26310

An intricate skeleton of lashed and fitted materials sits beneath an equally complex cover—a single kayak (*qajait*) requires ten or eleven sealskins to fit around its frame. Women prepare the skins, cut them to specifications, thread them together with watertight stitching, stretch the cover, and sew it onto the kayak. (Peters et al. 2002: 169; Zimmerly 2000: 55; Buijs 2004: 173)

**64**
Alaska, Kodiak Island
**Miniature three-person kayak (*baidarka*) frame**
Probably collected in the mid–19th century
13.1.593

The three-holed *baidarka* kayak remains popular in parts of Alaska today. Because they must withstand the added weight of passengers and gear, the frame is built from hemlock, a hard, crack-resistant wood. Vertical and horizontal supports between the openings, though not shown here, provide structural reinforcement. Keeping the kayak flexible required elastic straps of rope and bone, which connected many of the wooden joints in full-sized Aleut kayaks. (Zimmerly 2000: 35–36, 55)

The fifth section of the exhibition examines the perspectives of policy, science, and indigenous knowledge in observing and interpreting weather and the climate as all of them are brought together at conferences on climate change.

**65**

South Greenland

**Miniature of a hunting scene with blind and gun stand** (see p. 54)

Collected 1938

Wood, paper, cloth, leather, seal fur and leather, commercial thread, glue

Gift of George Murphy, Class of 1941; 39.70.7906

A seal, enjoying spring by basking at its breathing hole, has little hope of escape from this hunter, camouflaged in a white canvas parka, and his rifle, steadied by a gun stand. To get this close to the seal, the hunter must continually observe the seal's habits and move cautiously to remain "invisible." Many Inuit believe that seals consent to being hunted, but this does not make the hunt easy or the outcome certain.

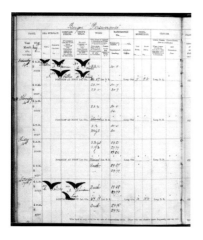

**66**

**Logbooks of the whaling ship** (see p. 14)

1877–80

Stefansson Collection on Polar Exploration, Rauner Special Collections Library, Dartmouth College Library; Stef MSS-121

For a whaling crew member in the nineteenth century, making logbook entries offered respite from hard work on deck while also creating valuable information about weather and hunting patterns. Today, these accounts, including Arctic ice cover, the seasonal activities of animals, and vegetation dynamics, provide a unique opportunity to compare contemporary and historical observations regarding Arctic environments. This logbook page from May 1879 documents a voyage from Peterhead, Scotland, to Baffin Island, Canada—one of the *Perseverance's* major whaling routes. Alongside values for wind and barometric pressure are sketches of single and double dorsal fins. A single whale fin represents a failed harpoon strike, but a double fin stands for success. Each denotes a captured whale.

# REFERENCES CITED

ACIA. 2004. *Climate Impact Assessment.* Scientific report. Cambridge: Cambridge University Press.

Adney, E. T., and H. I. Chapelle. 1964. *The Bark Canoes and Skin Boats of North America.* Washington, D.C.: Smithsonian Books.

AHDR. 2004. *AHDR: Arctic Human Development Report.* Akureyri: Stefansson Arctic Institute. 15–26.

Aporta, C., and E. Higgs. 2005. "Satellite Culture: Global Positioning Systems, Inuit Wayfinding, and the Need for a New Account of Technology." *Current Anthropology* 46(5): 729–53.

Birket-Smith, K. 1924. *Ethnography of the Egedesminde District with Aspects of the General Culture of West Greenland.* Meddelelser om Grønland 66. Copenhagen.

Boas, F. 1888. *The Central Eskimo.* Sixth Annual Report of the Bureau of American Ethnology, 1884–85. Washington, D.C.: Government Printing Office.

Bogoras, W. 1904. *The Chukchee.* The Jesup North Pacific Expedition 7. Leyden: A. J. Brill.

Brody, H. 1976. "Land Occupancy: Inuit Perceptions." In *Inuit Land Use and Occupancy Projects.* Ottawa: Department of Indian and Northern Affairs. 185–242.

Buijs, C. C. M. 2004. *Furs and Fabrics: Transformations, Clothing, and Identity in East Greenland.* Leiden: CNWS Publications.

Collignon, B. 1996. *Les Inuit, ce qu'ils savant du territoire.* Paris: L'Harmattan.

Fienup-Riordan, A. 2005. *Ciuliamta Akluit: Things of Our Ancestors.* Seattle: University of Washington Press.
———. 1996. *The Living Tradition of Yup'ik Masks.* Seattle: University of Washington Press.
———. 1990. "The Bird and the Bladder: The Cosmology of Central Yup'ik Seal Hunting." *Études/Inuit/Studies* 14(1–2): 23–28.

Fitzhugh, W. W., and S. A. Kaplan. 1982. *Inua: Spirit World of the Bering Sea Eskimo.* Washington, D.C.: Smithsonian Institution.

Garber, C. M. 1940. *Stories and Legends of the Bering Strait Eskimos.* Boston: Christopher Publishing House.

Government of Nunavut. January–March 2001. *Inuit qaujimajangit hilap alanguminganut / Inuit Knowledge of Climate Change: A Sample of Inuit Experiences of Climate Change in Nunavut: Baker Lake and Arviat, Nunavut.* Baker Lake and Arviat: Government of Nunavut, Department of Sustainable Development, Environmental Protection Services.

Hassol, S. J., ed. 2004. *ACIA: Impacts of a Warming Arctic: Arctic Climate Impact Assessment.* Cambridge: Cambridge University Press.

Heath, J. D., and E. Arima. 2004. *Eastern Arctic Kayaks: History, Design, Technique.* Fairbanks: University of Alaska Press.

Huntington, H., S. Fox, et al. 2005. "The Changing Arctic: Indigenous Perspectives." In S. J. Hassol, ed., *ACIA: Impacts of a Warming Arctic: Arctic Climate Impact Assessment.* Cambridge: Cambridge University Press. 61–98.

Issenman, B. 1997. *Sinews of Survival: The Living Legacy of Inuit Clothing.* Vancouver: University of British Columbia Press.

Kangok, K., M. Boki, and J. Shaimaiyuk. 2001. "The Spiritual Intervention of the Shaman in the Inuit World." In B. Saladin d'Anglure, ed., *Cosmology and Shamanism.* Iqaluit: Nunavut Arctic College. 220–28.

Kappianaq, G., and C. Nutaraq. 2001. *Traveling and Surviving on Our Land.* Iqaluit: Language and Culture Program of Nunavut Arctic College.

Kramvig, B. 2003. "Nature, Culture, Dreams, and Healing." In K. Pedersen and A. Viken, eds., *Nature and Identity: Essays on the Culture of Nature.* Kristiansand: Hoyskole Forlaget, 167–87.

Krupnik, I. 2002. "Watching Ice and Weather Our Way: Some Lessons from Yupik Observations of Sea Ice and Weather on St. Lawrence Island, Alaska." In I. Krupnik and D. Jolly, eds., *The Earth Is Moving Faster Now: Indigenous Observations of Arctic Environmental Change.* Fairbanks: Arctic Research Consortium of the United States. 156–97.

Krupnik, I., and D. Jolly, eds. 2002. *The Earth Is Moving Faster Now: Indigenous Observations of Arctic Environmental Change.* Fairbanks: Arctic Research Consortium of the United States.

Lantis, M. 1959. *Folk Medicine and Hygiene: Lower Kuskokwim and Nunivak-Nelson Island Areas.* Anthropological Papers of the University of Alaska 8(1).
———. 1947. *Alaskan Eskimo Ceremonialism.* American Ethnological Society Monograph 11. New York: J. J. Augustin.

Laugrand, F. 2002a. "Éscrire pour prendre la parole: Conscience historique, mémoires d'aînés et régimes d'historicité au Nunavut." *Anthropologie et Sociétés* 26 (2–3): 91–116.
———. 2002b. *Mourir et renaître: La réception du christianisme par les Inuit de l'Arctique de l'Ést canadien (1890–1940).* Leiden: Research School CNWS.

MacDonald, J. 1998. *The Arctic Sky: Inuit Astronomy, Star Lore, and Legend.* Iqaluit: Royal Ontario Museum, Nunavut Research Institute, Toronto.

Nunavut Land Claims Agreement. 1993. *Agreement between the Inuit of the Nunavut Settlement Area and Her Majesty the Queen in Rights of Canada.* Ottawa: Department for Indian and Northern Affairs.

Nuttall, M. 1992. *Arctic Homeland: Kinship, Community, and Development in Northwest Greenland.* Toronto: University of Toronto Press.

Nuttall, M., F. Berket, B. Forbes, G. Kofinas, and G. Wenzel. 2005. "Hunting, Herding, Fishing, and Gathering: Indigenous Peoples and Renewable Resource Use in the Arctic." In *ACIA: Arctic Climate Impact Assessment.* Cambridge: Cambridge University Press.

Omura, K. 2002. "Construction of *Inuinnaqtun* (Real Inuit-Way): Self-Image and Everyday Practices in Inuit Society." *Self- and Other-Images of Hunter-Gatherers.* Paper presented at the 18th International Conference on Hunting and Gathering Societies (CHAGS 8), National Museum of Ethnology, October 1998. Osaka. 101–11.

Oosten, J. 1989. "Theoretical Problems in the Study of Inuit Shamanism." In M. Hoppal and O. J. von Sadovsky, eds., *Shamanism: Past and Present.* Budapest: Fullerton. 331–41.

Oosten, J., F. Laugrand, and M. Kakkik. 2003. *Keeping the Faith.* Memory and History in Nunavut 3. Iqaluit: Nunavut Arctic College.

Oosten, J., and C. Remie. 1999. Introduction. In J. Oosten and C. Remie, eds., *Arctic Identities: Continuity and Change in Inuit and Saami Societies.* Leiden: Research School CNWS. 1–4.

Oozeva, C., C. Noongwook, G. Noongwook, C. Alowa, and I. Krupnik. 2004. *Watching Ice and Weather Our Way / Sikumengllu eslamengllu esghapalleghput.* Washington D.C.: Arctic Studies Center, Smithsonian Institution.

Pelly, D. 2001. *Sacred Hunt: A Portrait of the Relationship between Seals and Inuit.* Vancouver: Greystone Books.

Peter, A., M. Ishulutak, J. Shaimaiyuk, N. Kisa, B. Kootoo, and S. Enuaraq. 2002. "The Seal: An Integral Part of Our Culture." *Études/Inuit/Studies* 1(26): 167–74.

Peter, A., and N. Kisa. 2001. "The Inuit Life Cycle." In B. Saladin d'Anglure, ed., *Cosmology and Shamanism.* Iqaluit: Nunavut Arctic College. 199–207.

Rasmussen, K. 1931. *The Netsilik Eskimo: Social Life and Spiritual Culture.* Report of the Fifth Thule Expedition, 1921–24, vol. 8. Copenhagen: Nordisk.
———. 1930. *Intellectual Culture of the Caribou Eskimos.* Report of the Fifth Thule Expedition, 1921–24, vol. 7(2). Copenhagen.
———. 1929. *Intellectual Culture of the Iglulik Eskimos.* Report of the Fifth Thule Expedition, 1921–24, vol. 7(1), Nordisk, Copenhagen.
———. 1927. *Across Arctic America: Narrative of the Fifth Thule Expedition.* New York: G. P. Putnam's Sons.

Rousselot, J. 1994 *Kanuitpit? Kunst und Kulturen der Eskimo; eine Auswahl aus den Museumssammlungen.* Munich: Staatliches Museum fur Völkerkunde.

Routledge, M., and M. E. Jackson. 1991. *Pudlo: Thirty Years of Drawing.* Ottawa: National Gallery of Canada.

Saladin d'Anglure, B. 1990. "Frère-lune (Taqqiq), soeur-soleil (Siqiniq), et l'intelligence du monde (Sila)."*Études/Inuit/Studies* 14(1–2): 75–139.
———. 1986. "Du foetus au chamane: La construction d'un 'troisième sexe' Inuit." *Études/Inuit/Studies* 10(1–2): 25–113.
———. 1980. "'Petit-Ventre.' l'enfant-géant du cosmos Inuit: Ethnographie de l'enfant et enfance de l'ethnographie dans l'Arctique central Inuit." *L'homme* 20(1): 7–46.

Stevenson, M. G. 1997. *Inuit, Whalers, and Cultural Persistence: Structure in Cumberland Sound and Central Inuit Social Organization.* New York: Oxford University Press.

Stuckenberger, A. N. 2005. *Community at Play: Social and Religious Dynamics in the Modern Inuit Community of Qikiqtarjuaq.* Amsterdam: Rozenberg Publishers.

Sturtevant, W. C., and D. Damas. 1984. *Arctic.* Handbook of North American Indians, vol. 5. Washington, D.C.: Smithsonian Institution.

Thorbe, N., S. Eyegetok, and Hakongak, Naika, Kitikmeot Elders. 2002. "Nowadays It Is Not the Same: Inuit Qujimajauqangit, Climate, and Caribou in the Kitikmeot Region of Nunavut, Canada." In I. Krupnik and D. Jolly, eds., *The Earth Is Moving Faster Now: Indigenous Observations of Arctic Environmental Change.* Fairbanks: Arctic Research Consortium of the United States. 198–239.

Varjola, P. 1990. *The Etholen Collection.* Helsinki: National Board of Antiquities of Finland.

Weyer, M. 1932. *The Eskimos: Their Environment and Folkways.* New Haven: Yale University Press.

Woodward, K. E. No date. "Northern Art and Ethnographic Material in the Collections of Dartmouth College." *Northern Lights.* Hanover, N.H.: Trustees of Dartmouth College.

Working Group on Traditional Knowledge. 1998. *Presentations.* Iqaluit: Working Group on Traditional Knowledge.

Young, O. R., and N. Einarsson. 2004. "Introduction: Human Development in the Arctic." *AHDR: Arctic Human Development Report.* Akureyri: Stefansson Arctic Institute. 15–26.

Zimmerly, D. 2000. *Qayak: Kayaks of Alaska and Siberia.* 2nd ed. Fairbanks: University of Alaska Press.

## WEB PAGES CONSULTED

http://alaska.fws.gov/fisheries/mmm/walrus/nhistory.htm

http://coas.missouri.edu/anthromuseum/pdfs/l_huntfish.pdf

http://pwnhc.learnnet.nt.ca/exhibits/nv/harpoon.htm

http://www.acia.uaf.edu

http://www.civilization.ca/aborig/watercraft/wak04eng.html

http://www.dfo-mpo.gc.ca/zone/underwater_sous-marin/hseal/seal-phoque_e.htm

http://www.gov.nu.ca/Nunavut/English/departments/DSD/

http://www.pc.gc.ca/canada/nature/archives/2003/av-ap/archives1_E.asp

http://scaa.usask.ca/gallery/northern/content?pg=ex13-3

http://www.tribalarts.com/feature/riordan/

http://www2.nature.nps.gov/geology/parks/kova/index.cfm

http://www.gov.nu.ca/Nunavut/English/departments/DSD/

http://forces.si.edu/arctic.html

# CONTRIBUTORS

**A. Nicole Stuckenberger** curated *Thin Ice: Inuit Traditions within a Changing Environment* at the Hood Museum of Art, Dartmouth College. The exhibition is a part of Dartmouth's contributions to the International Polar Year 2007–8. Dr. Stuckenberger received her doctorate in cultural anthropology from Utrecht University in 2005. She is the author of "Community at Play: Social and Religious Dynamics in the Modern Inuit Community of Qikiqtarjuaq," the result of fourteen months of fieldwork with the Inuit of Qikiqtarjuaq, Nunavut (Baffin Island, Canada). Dr. Stuckenberger is the current Stefansson Postdoctoral Fellow at the Institute of Arctic Studies, John Sloan Dickey Center for International Understanding, Dartmouth College.

**William Fitzhugh** is an anthropologist specializing in circumpolar archaeology, ethnology, and environmental studies. He received is B.A. from Dartmouth in 1964, where he studied anthropology with Professor Elmer Harp Jr. He first became interested in the North through his work with Professor Harp, who invited him to take part in archaeological projects in Newfoundland and Hudson Bay. After graduating from Dartmouth, he spent two years in the U.S. Navy, followed by graduate studies at Harvard University. He received his Ph.D. in anthropology in 1970 and thereafter took a position at the National Museum of Natural History. As director of the Arctic Studies Center and Curator in the Department of Anthropology there, he has spent more than thirty years studying and publishing on Arctic peoples and cultures in northern Canada, Alaska, Siberia, and Scandinavia. His archaeological and environmental research has focused upon the prehistory and paleo-ecology of northeastern North America, and broader aspects of his research feature the evolution of northern maritime adaptations, circumpolar culture contacts, cross-cultural studies, and acculturation processes in the North, especially concerning Native-European contacts. As curator of the National Museum of Natural History's Arctic collections, Dr. Fitzhugh has produced four international exhibitions, *Inua: Spirit World of the Bering Sea Eskimos; Crossroads of Continents: Native Cultures of Siberia and Alaska; Ainu: Spirit of a Northern People;* and *Vikings: The North Atlantic Saga.* His public and educational activities include the production of films including the NOVA specials *Mysteries of the Lost Red Paint People, Norse America,* and several other Viking films. He is an advisor to the Arctic Research Commission, represents the Smithsonian and Arctic social sciences in various inter-agency councils, serves on the Smithsonian Science Commission, and holds various other administrative and advisory posts.

**Aqqaluk Lynge** represented the Inuit of Alaska, Canada, Greenland, and the Far East of Russia as president of the Inuit Circumpolar Conference (ICC) from 1997 to 2002. At the ICC's 9th General Assembly in Kuujjuaq, 2002, Mr. Lynge was appointed President of ICC Greenland and ICC Vice-Chair. Mr. Lynge started his professional career as a social worker after graduating from the National Danish School of Social Work in 1976. He was a radio broadcaster for several years before entering Greenland politics. He has promoted the rights of Indigenous Peoples both in his home country of Greenland and internationally since his youth. He has also demonstrated a deep commitment to pan-Inuit unity since the early 1970s and, before becoming ICC president in 1997, served as a continuous member of the ICC Executive Council beginning in 1980. Mr. Lynge was first elected to the Greenland Parliament in 1983 and has served both as a member of Parliament and as a minister of various portfolios. Mr. Lynge is widely published, having written books of poetry, essays, and politics. He has also contributed to several works and anthologies written in the English, Greenlandic, French, and Nordic languages. He has been an invited speaker at a wide range of international human rights fora, at wildlife management conferences, environmental summits, Arctic Council Ministers' summits, and others. Most recently, he has been a very active supporter of the Thule Case, which dealt with the 1953 forced relocation of Greenlandic Inuit to make way for the U.S. military base in northern Greenland.

**Kesler Woodward,** an Alaska resident since 1977, served as Curator of Visual Arts at the Alaska State Museum and as Artistic Director of the Visual Arts Center of Alaska before moving to Fairbanks in 1981. At Dartmouth College, he was a Visiting Research Fellow and Director of Arts and Humanities Projects for the Institute on Canada and the United States from 1989 through 1991. He is currently Professor of Art Emeritus at the University of Alaska, Fairbanks, where he taught for two decades, serving as Chair of the Art Department and as Chair of the Division of Arts and Communications. He retired from teaching to paint full time in the spring of 2000. Woodward's paintings are included in all major public art collections in Alaska, and in museum, corporate, and private collections on both coasts of the United States. He has had solo museum exhibitions at the Morris Museum of Art, the University of Alaska Museum, the Alaska State Museum, and the Anchorage Museum of History and Art. Also an art historian and curator, Woodward has published five books since 1990 on Alaskan art, including the first comprehensive survey of the fine arts in Alaska, *Painting in the North,* published by the Anchorage Museum and University of Washington Press in 1993. His latest volume, *Painting Alaska,* was published by the Alaska Geographic Society in 2000. He has lectured on the art of the circumpolar north from Alaska to Georgia, New England, and the British Museum in London.